GMP and Quality Handbook

An Introduction

Beverly White

ISBN: 9798882160141
Imprint: Independently published

Contents

1.0   QUALITY MANAGEMENT SYSTEMS

1.1 Introduction

Quality management systems are designed to provide a framework where multiple processes, requirements, inputs and outputs work in a coordinated fashion that ensure overall effective systems are maintained over time. Numerous subsystems form part of a quality management system. Each system has procedures and written instructions that outline the requirements for that functional area. For example a QMS shall have a system for handling customer complaints, and a system for handling nonconformances, a system for handling validation activities, a system for handling change activities and so on.

Each subsystem may also have inputs in the form of information or numerical data which is assessed and processed in order to make decisions or recommendations. A simple example of inputs would be customer complaints or adverse events that feed into complaint management and post-market surveillance activities. The quality management is an ecosystem of processes that are interrelated that work to provide effective management with continuous improvement at its heart.

Several authorities around the world mandate quality management systems as a requirement for the manufacture, registration and sale of medical devices. These requirements are laid out in legislation that must be complied with by manufacturers or organisations involved in medical device activities. Such regulations include FDA 21 CFR part 820 which is the code of federal regulations which apply to medical devices in the United states.

Within general manufacturing and services industries the ISO 9001 quality management system Is utilised to deliver consistent products or services high quality products and services which benefit the business also. more specifically for medical devices also thirteen 485 quality management system also targets principles of quality products customer satisfaction but go further in order to deliver product's dot are safe and effective as medical devices some core principles are common between ISO 9001 and also 13485.

The pharmaceutical industry plays a pivotal role in promoting global health by developing and manufacturing essential medicines. However, the efficacy and safety of pharmaceutical products are contingent upon stringent manufacturing practices. Good Manufacturing Practice (GMP) serves as the bedrock of quality assurance in pharmaceutical manufacturing, encompassing a comprehensive set of guidelines and regulations. Compliance with GMP requirements is imperative to ensure that pharmaceutical products consistently meet quality standards, minimizing risks to patients and maintaining public confidence in the industry. This essay explores the essential aspects of GMP requirements for pharmaceuticals, drawing insights from regulatory authorities such as the FDA, EMA, and WHO.

1.2 ISO Quality Management System for Medical Devices

ISO 13485 -Medical Devices, Quality Management system is an internal standard for use within medical device manufacturing companies and organizations involved in the design and/or development, production, storage and distribution, installation, or servicing of a medical device and design and development or provision of technical or professional services. It is a standard that applies throughout the product life-cycle.

The recent revision is designed to address recent developments in quality management and other updated regulations that relate to the industry. Improvements in the new version of the standard include broadening its applicability to include all organizations involved in the lifecycle of the product, from the concept stage to end of life along with greater alignment with regulatory requirements and post-market surveillance including complaint handling. ISO 13485 is also used by suppliers or external vendors that provide QMS related services. It has largely become an essential element for manufacturers placing product on the market throughout the world. While it differs to regulations in so far it is an international standard, it has become integrated into many national competent authorities requirements for market approval and certification activities. Requirements within the standard are applicable regardless of the size or an organisations. It should be noted that if clause 6, 7 or 8 of ISO 13485 is not applicable due to the activities undertaken by the organisation or the nature of the medical device for which the quality management system is applied, the organisation does not need to include such a requirement in its quality management system. It uses a process approach to quality management which is made up of subprocesses where inputs are used and outputs inform other processes.

1.3 Medical Devices, Code of Federal Regulations, FDA, 21 CFR Part 820

It is necessary for manufacturers established a quality (management) system if they intend to market commercial medical device products in the United states. The requirement is concise in nature but it has wide implications for a the manufacturer. In establishing a QMS there needs to be broad commitment within the organization and the resources, training and personnel provided to achieve implementation and in ongoing application and maintenance of the quality management system.

> *Reference: 820.5 Quality system.*

> *'Each manufacturer shall establish and maintain a quality system that is appropriate for the specific medical device(s) designed or manufactured, and that meets the requirements of this part.'*

21 CFR Part 820 consists of 15 subparts:

Subpart A - General Provisions
Subpart B - Quality System Requirements
Subpart C - Design Controls
Subpart D - Document Controls

Subpart E - Purchasing Controls
Subpart F - Identification and Traceability
Subpart G - Production and Process Controls
Subpart H - Acceptance Activities
Subpart I - Nonconforming Product
Subpart J - Corrective and Preventive Action
Subpart K - Labelling and Packaging Control
Subpart L - Handling, Storage, Distribution, and Installation
Subpart M - Records
Subpart N - Servicing
Subpart O - Statistical Techniques

1.4 Code of Federal Regulations, FDA, 21 CFR Part 211

The FDA, as the regulatory authority responsible for ensuring the safety, efficacy, and quality of pharmaceutical products in the United States, promulgates GMP regulations under Title 21 of the Code of Federal Regulations (CFR), specifically in Parts 210 and 211. These regulations, commonly referred to as the Current Good Manufacturing Practice (cGMP) regulations, outline the minimum requirements for the manufacture, processing, packing, or holding of drugs. Key provisions of cGMP FDA regulations for pharmaceuticals includes:

- Personnel Qualifications and Training: Requirements for personnel training, hygiene, and health status to ensure competency and prevent contamination.

- Facility and Equipment Standards: Specifications for pharmaceutical facilities, utilities, and equipment to maintain appropriate environmental conditions and prevent cross-contamination.

- Documentation and Record-keeping: Mandates for comprehensive documentation of manufacturing activities, including batch records, standard operating procedures, and quality control records.

- Quality Control Testing: Requirements for in-process and finished product testing to verify compliance with specifications and ensure product quality and safety.
- Process Validation: Guidelines for validating manufacturing processes to demonstrate consistency, reproducibility, and control over critical parameters.

The FDA conducts inspections of pharmaceutical manufacturing facilities to assess compliance with cGMP regulations, issuing warning letters, Form 483 observations, or enforcement actions in cases of non-compliance.

1.5 European Medicines Agency, (EMA):

In the European Union (EU), the EMA serves as the central regulatory authority responsible for the evaluation and supervision of medicinal products. GMP requirements for pharmaceuticals and medicinal products in the EU are legislated at European level in Directive 2003/94/EC and its annexes, as well as in Volume 4 of the EudraLex guidelines. Key aspects of GMP requirements in the EU include:

- Authorization and Inspection: Requirement for pharmaceutical manufacturers to obtain a manufacturing authorization from competent authorities and undergo regular inspections to ensure compliance with GMP standards.
- Quality Management System: Implementation of a comprehensive quality management system encompassing personnel, premises, equipment, documentation, and quality control measures.
- Product Quality and Testing: Emphasis on product quality attributes, specifications, and testing methods to ensure consistency, safety, and efficacy throughout the product lifecycle.
- Documentation and Record-keeping: Obligation to maintain detailed documentation of manufacturing processes, including batch records, standard operating procedures, and quality control records, in accordance with regulatory requirements.
- Outsourced Activities: Requirements for pharmaceutical companies to ensure oversight and control over outsourced activities, including contract manufacturing, testing, and distribution.

1.5.1 Product Quality

Product quality is a fundamental aspect of pharmaceutical manufacturing, encompassing attributes such as safety, efficacy, purity, and consistency. Ensuring the quality of pharmaceutical products is imperative to safeguard patient health and confidence in the industry. Regulatory authorities, including the European Union (EU), have established comprehensive guidelines to govern pharmaceutical quality throughout the product lifecycle. Among these guidelines, EU Volume 4 holds particular significance, offering detailed requirements and recommendations for achieving and maintaining product quality in accordance with Good Manufacturing Practice (GMP) standards. This essay explores the essential principles of product quality in pharmaceuticals and examines the key provisions of EU Volume 4 guidelines, highlighting their importance in promoting excellence and innovation in pharmaceutical manufacturing. Product quality is a multifaceted concept that encompasses various attributes essential to the safety, efficacy, and reliability of pharmaceutical products. In the context of pharmaceutical manufacturing, ensuring product quality is paramount for several reasons described below.

1.5.2 Patient Safety

High-quality pharmaceutical products are essential for protecting patient safety and well-being. Substandard or adulterated medicines can pose significant risks to patients, ranging from adverse reactions to therapeutic failure or even life-threatening consequences.

1.5.3 Therapeutic Efficacy

The effectiveness of pharmaceutical treatments relies on the quality and consistency of active pharmaceutical ingredients (APIs) and excipients. Product quality ensures that medicines deliver the intended therapeutic benefits and achieve desired clinical outcomes for patients.

1.5.4 Regulatory Compliance

Regulatory authorities enforce stringent quality standards to ensure that pharmaceutical manufacturers adhere to Good Manufacturing Practice (GMP) guidelines. Compliance with regulatory requirements is essential for obtaining marketing authorization and maintaining licensure to market pharmaceutical products.

1.5.5 Summary of EU Volume 4 for Pharmaceutical Quality

The European Union (EU) Volume 4 guidelines, part of the European Medicines Agency (EMA) regulatory framework, provide detailed guidance on pharmaceutical quality assurance throughout the product lifecycle. Volume 4 of the EudraLex guidelines encompasses Annex 1 to Annex 20, covering a wide range of topics related to pharmaceutical manufacturing, quality control, and regulatory compliance. Key aspects of EU Volume 4 guidelines include:

1. Quality Management System (QMS):

The foundation of EU Volume 4 guidelines lies in the establishment and maintenance of a robust Quality Management System (QMS) within pharmaceutical manufacturing facilities. A QMS encompasses organizational structure, responsibilities, processes, procedures, and resources dedicated to ensuring product quality and regulatory compliance. Key components of a QMS include:

- Quality Policy: A statement of the organization's commitment to quality, outlining objectives, responsibilities, and expectations for achieving and maintaining product quality.
- Quality Risk Management: The systematic process of identifying, assessing, controlling, and mitigating risks to product quality throughout the product lifecycle. Risk management principles guide decision-making and resource allocation to prioritize quality objectives and minimize potential hazards.
- Quality Assurance: The implementation of policies, procedures, and systems to ensure that pharmaceutical products consistently meet established quality standards and regulatory requirements. Quality assurance activities encompass document control, change management, deviation management, and batch release.

- Quality Control: The testing, analysis, and evaluation of raw materials, in-process samples, and finished products to verify compliance with specifications and quality attributes. Quality control measures include analytical testing, stability testing, microbiological testing, and environmental monitoring.
- Continuous Improvement: A commitment to ongoing evaluation, optimization, and enhancement of manufacturing processes, quality systems, and operational practices to achieve and sustain excellence in product quality.

2. Process Validation:

Process validation is a key component of EU Volume 4 guidelines, ensuring that pharmaceutical manufacturing processes are designed, qualified, and operated to consistently produce products of the desired quality. Process validation involves three stages:

- Process Design: The initial stage of process validation involves defining the manufacturing process, identifying critical process parameters (CPPs), and establishing control strategies to ensure product quality and reproducibility.
- Process Qualification: The second stage of process validation entails demonstrating the capability of the manufacturing process to consistently produce products that meet predefined quality attributes within specified operating ranges. Process qualification activities include installation qualification (IQ), operational qualification (OQ), and performance qualification (PQ).
- Continued Process Verification: The third stage of process validation involves ongoing monitoring and verification of the manufacturing process to ensure continued compliance with quality standards and regulatory requirements. Continued process verification activities include process performance monitoring, trending analysis, and periodic revalidation.

Process validation is essential for ensuring the robustness, reliability, and reproducibility of pharmaceutical manufacturing processes, thereby minimizing the risk of product variability or non-conformance.

3. Quality Control Testing:

Quality control testing plays a critical role in EU Volume 4 guidelines, enabling pharmaceutical manufacturers to verify the quality, purity, potency, and safety of raw materials, intermediates, and finished products. Quality control testing encompasses various analytical techniques and methodologies, including:

- Chemical Analysis: Identification and quantification of active pharmaceutical ingredients (APIs), excipients, and impurities using chromatographic techniques such as high-performance liquid chromatography (HPLC), gas chromatography (GC), and mass spectrometry (MS).
- Physical Testing: Evaluation of physical properties such as particle size, density, viscosity, dissolution rate, and friability to assess product quality, uniformity, and performance.

- Microbiological Testing: Detection and enumeration of microorganisms, including bacteria, fungi, and viruses, to ensure the sterility, microbial purity, and safety of pharmaceutical products.
- Stability Testing: Evaluation of product stability under various environmental conditions, including temperature, humidity, and light exposure, to assess shelf-life, storage conditions, and degradation pathways.

The EMA collaborates with national competent authorities within EU member states to harmonize GMP standards and facilitate mutual recognition of manufacturing authorizations.

1.6 Quality Management System Overview

The structure of a Quality management system can be documented in a Quality Policy. This is a controlled document that sets out all the elements of the QMS within a company or organization. As such it is one of the primary QMS documents that are called upon during audits. The Quality Policy is then implemented and supported by procedures and processes addressing key functional areas of the QMS used to facilitate manufacturing or medical device regulated activities such as design, procurement, distribution, contract services and so on.

Inputs and outputs of each element result in an integrated process approach to quality management. The input/output feature allows continual review and responsiveness of the QMS at a sub process level. To illustrate this; management are responsible for the QMS been established, implemented and maintained. The performance of the quality management system can be evaluated based on multiple factors (nonconformances, CAPA, audit). Risk Management is another factor that can be considered in the overall effectiveness of the system. If deficiencies in the risk management process or the risk profile of the product changes adversely based on product performance in the field, then this can indicate deeper issues that may be inherent in the system.

A Quality Policy must also set out the organizations commitment to quality. Commitment is a fundamental principle which is intended to ensure quality is an ongoing theme for the company and not just a once off activity. Commitment must be demonstrated from top management. Often a signed and approved quality policy is displayed by companies in its reception of hosting area.

The application of ISO 13485 as a Quality management standard provides companies with a structured approach to establishing, implementing and maintaining a QMS. Compliance to it works to satisfy regulatory requirements where a QMS fundamental requirement.

Management Responsibility

This element of quality management covers Management commitment, Customer focus, Quality policy, Planning, Responsibility, authority and communication and Management review.

The management are responsible for "Establishing a framework under which quality requirements are identified and structured into a functional organization that supports both customer and regulatory requirements." As part of the quality management system, risk must be managed throughout the organisation and the necessary processes and controls are in place to monitor maintain the systems. A Management Representative is assigned by top management and is responsible for ensuring that processes are established, implemented and maintained. The processes and systems allow Management provide leadership and oversight and facilitates communication required for the business. Reporting to top management regularly on the performance and effectiveness of the quality management system with any needs for improvement is also a responsibility of the management representative.

Management Commitment

It is essential that Top Management have a commitment to meet regulatory requirements and reach a high level of product quality to meet the expectations of customers and provide safe and effective products.. Top Management commitment is demonstrated by the implementation, maintenance and continued effectiveness of the quality management system. Communication must be at the heart of a successful QMS, starting with Top Management with communication internally and to regulators and customers.

Quality Policy

Each company must have a written Quality Policy as a requirement of an ISO 13485 QMS. And FDA Q.S. A Quality Policy has several purposes that an organization must cover. The policy sets out the organizations commitment to quality, management commitment a fundamental principle which is intended to ensure quality is an ongoing priority. The policy should also summarize the overall intentions and directions of an organization with respect to quality. The policy should touch upon customer requirements, regulatory requirements and set out the core objectives and goals of the company. A clear quality policy with realistic objectives is a requirement of management commitment and is the starting point of any Quality management system. The objectives and goals within the Quality policy should be supported and reflected by the content and processes in methods, procedures and documentation throughout the organization. The goals and objectives should also be suitable to the needs and type of business.

Customer Focus

Customer feedback is a requirement of ISO 13485 and as such the manufacturer must engage with the customer. In instances where a defective product is received, the manufacturer must have a complaints process to facilitate proper feedback, communication and investigation.

Management Review

Each company or manufacturer needs to have written procedures in place for conducting management reviews and to describe the activities that support management review and provide input -e.g. quality audits. Procedures should make reference to defined intervals or a schedule of management review and audits.

Management reviews are important as they ensure top management are involved and supporting the quality management system. Management reviews are also a formal requirement of ISO 13485 ensuring that the quality system is effective and is given the right amount of attention to sustain its effectiveness.

Change Management

Management of Changes are required for (1) changes to the quality management system itself
(2) changes made to products including raw materials, sub-components, materials, labels and other supplies intended for commercial sale or clinical investigation.

Regulatory Management

Regulatory affairs covers a number of regulatory management requirements. For marketed products, the technical file for products and markets needs to be maintained in an accurate and compliant manner. They also are essential in providing regulatory guidance and expertise for new products and new registrations.

Design Controls

The Federal Drug Administration (FDA) regulations, 21 CFR Part 820.30, subpart C - Design Controls provides the basis for product development and product realisation for manufacturers. Its content includes section on: (a) General, (b) Design and development planning, (c) Design input, (d) Design output, (e) Design review, (f) Design verification, (g) Design validation, (h) Design transfer, (i) Design changes, (j) Design history file. These requirements are covered in the chapter on Product Realization.

ISO 13485, covers product realisation under the following sub clauses-*Planning of product realization 7.1*
Customer-related processes 7.2, Design and development 7.3, Purchasing 7.4, Production and service provision 7.5
And Control of monitoring and measuring equipment 7.6.
Both approaches (21 CFR 820.30 & ISO 13485) ensure products are designed and developed in a controlled fashion, with adequate documentation and management review and oversight over the course of development and on into commercialisation.

Inspection of Quality Management Systems

The Quality System Inspection Technique adopted by the FDA uses the "establish" approach in conducting inspections. For each subsystem, an auditor will first determine if the company has defined and documented the requirements (CAPA, Design, etc.) by looking at procedures and policies.

Inspection Objectives

Verify the Quality Policy and Objectives have been implemented:

- Quality Policy
- Management Review
- Quality Audit procedures
- Quality Plan
- Quality Management System Procedures

Verify the Quality Policy and Objectives have been implemented:

- and is applicable to the purpose of the organization
- includes a commitment to comply with:
  - QMS requirements and to
  - maintain the effectiveness of the QMS
- method of establishing and reviewing quality objectives
  - Policy is communicated and understood within the organization
  - Policy is reviewed to ensure it reviewed for suitability

Subsystems or subprocesses form the foundations of a quality management system. A subsystem inspection approach focuses on the important and critical aspects of quality system regulation.

The process for performing subsystem inspections is based on a "top-down" approach to inspecting and is typical of competent authorities and notified bodies worldwide. Both "top-down" and "bottom-up" inspectional approaches involve record reviews. A "top-down" approach to audits and inspections involves looking at the "systems" for addressing quality compliance prior to the examination of specific quality problems or issues. A bottom-up approach to inspection involves looking a individual quality issues or non-conformances and then work back up through the quality system. The advantage of this methodology is that it allows the auditor to focus on specific problems followed by evaluating the response & actions in relation to that issue.

Quality System Procedures and Instructions

Manufacturers must prepare and implement all activities, including, but not necessarily limited to the applicable requirements of the Quality System Regulation (or Quality Management System), that are necessary to assure the finished product meets all pre-determined specifications. The "quality system" as specified in the FDA 21 CFR, Quality System Regulation refers to all activities previously referred to as "quality assurance". Slightly different terms may be used by various manufacturers such as "quality control" or "GMP Control" or "quality assurance" instead of quality system. However, this is acceptable once all the provisions of the requirements are satisfied.

1.7 Good Manufacturing Practices (GMP)

Good Manufacturing Practices are a set of practices that are required in order to comply with industry standards and regulations. GMP helps to minimise the risks involved during manufacturing and helps to ensure products meet quality and regulatory standards. A GMP quality system ensures that products are consistently produced and controlled according to predefined quality standards. It is designed to minimise the risks involved in any pharmaceutical production that cannot be eliminated through testing the final product. Often, a broader term is used in industry -GxP-where the "x" is used as an umbrella letter representing different subjects or disciplines in industry. Some prime examples include GLP (Good Laboratory Practice), GDP (Good Documentation Practice), GEP (Good Engineering Practice) and GMP (Good Manufacturing Practices). Furthermore, the use of a lower case "c" as a prefix indicates "current" or "up-to-date". So cGMP stands for "Current Good Manufacturing Practices. This means that some conventions or practices are subject to change within the industry. Therefore, it is important to be up-to-date in the application of cGxP or cGMP

There are multiple regulators and organisations that provide definitions of "Good Manufacturing Practices". They include Organisations such as the World Health Organisation (WHO) and the International Society of Pharmaceutical Engineering (ISPE), PIC/s, EU Eurdralex Volume 4, Good Manufacturing Pracatices Other definitions are offered by bodies such as the American competent authority for Food and Drug Administration. It is good to have an awareness of how organisations, bodies and competent authorities define GMP, and one should always review the "local" regulatory landscape. Below some definitions are provided to provide a feel for GMP and highlight the common thread between definitions.

W.H.O. World Health Organisation-"Good Manufacturing Practices (GMP, also referred to as 'cGMP' or 'current Good Manufacturing Practice') is the aspect of quality assurance that ensures that medicinal products are consistently produced and controlled to the quality standards appropriate to their intended use and as required by the product specification."

Food and Drug Administration: cGMP refers to the Current Good Manufacturing Practice regulations enforced by the US Food and Drug Administration (FDA). cGMPs ensure systems are properly designed and monitored, safeguarding the control of manufacturing processes and facilities. Adherence to the cGMP regulations ensures the identity, strength, quality, and purity of drug products by requiring that manufacturers of medications adequately control manufacturing operations. This includes establishing strong Quality Management Systems, obtaining appropriate quality raw materials, establishing robust operating procedures, detecting and investigating product quality deviations and maintaining reliable testing laboratories. This formal system of controls at a pharmaceutical company, if adequately put into practice, helps to prevent instances of contamination, mix-ups, deviations, failures and errors. This assures that drug products meet their quality standards.

MHRA (Medicines and Healthcare Products Regulatory Agency) defines GMP as follows:

"Good Manufacturing Practice (GMP) is that part of quality assurance which ensures that medicinal products are consistently produced and controlled to the quality standards appropriate to their intended use and as required by the marketing authorisation (MA) or product specification. GMP is concerned with both production and quality control. Many of the drivers of GMP in effect are also benefits to the manufacturer. Good manufacturing practices are an expected practice in regulated industries and a manufacturer must meet all relevant GMP regulations if they wish to manufacture within a country or sell to a particular market. It is important to maintain accurate, complete, up-to-date and consistent information to ensure patient safety and reduce any potential risks."

A basic tenet of GMP is that (1) quality cannot be tested into a batch of product and (2) quality must be built into each batch of product during all stages of the manufacturing process. Good Manufacturing Practice (GMP) describes a set of principles and procedures that when followed helps ensure that therapeutic goods are of high quality. There are different codes of GMP, depending on the type of therapeutic good:

- ➢ Good Manufacturing Practice for Medicines
- ➢ Good Manufacturing Practice for Human Blood and Tissues
- ➢ A different system, known as conformity assessment, is used to ensure that medical devices are of high quality.

PICS/s manufacturing principles for medicinal products:

Pharmaceutical Inspection Convention and Pharmaceutical Inspection Co-operation Scheme (PIC/S): The Pharmaceutical Inspection Convention and Pharmaceutical Inspection Co-operation Scheme (jointly known as PIC/S) develop international standards between countries and pharmaceutical inspection authorities, to provide a harmonised and constructive co-operation in the field of Good Manufacturing Practices.The PIC/S provides an active and constructive cooperation in the field of GMP and related areas. The purpose of PIC/S is to facilitate:

- ➢ networking between participating authorities
- ➢ maintenance of mutual confidence

> ➢ exchange of information and experience
> ➢ Mutual training of GMP inspectors.

The Guide consists of an Introduction section along with two parts and a number of annexes.

- **Guide to Good Manufacturing Practice for Medicinal Products** – Introduction

  o Introduction
  o Adoption and entry into force
  o Revision history

- **Guide to Good Manufacturing Practice for Medicinal Products** - Part I
  Part I covers GMP principles for the manufacture of medicinal products

  1. Quality management
  2. Personnel
  3. Premises and equipment
  4. Documentation
  5. Production
  6. Quality control
  7. Contract manufacture and analysis
  8. Complaints and product recall
  9. Self-inspection

- **Guide to Good Manufacturing Practice for Medicinal Products** - Part II
  Part II covers GMP for active substances used as starting materials

  1. Introduction
  2. Quality management
  3. Personnel
  4. Buildings and facilities
  5. Process equipment
  6. Documentation and records
  7. Materials management
  8. Production and in-process controls
  9. Packaging and identification labelling of APIs and intermediates
  10. Storage and distribution
  11. Laboratory controls
  12. Validation
  13. Change control
  14. Rejection and re-use of materials
  15. Complaints and recalls
  16. Contract manufacturers (including laboratories)
  17. Agents, brokers, traders, distributors, repackers and relabellers
  18. Specific guidance for APIs manufactured by cell culture / fermentation
  19. APIs for use in clinical trials
  20. Glossary

  The annexes provide detail on specific areas of activity and are listed below:

- **Technical interpretation of PIC/S GMP guide Annex 1 -** Manufacture of sterile medicinal products. PIC/S has published a recommendation for the technical interpretation of Annex 1 on the Manufacture of Sterile Medicinal Products. This recommendation summarises the interpretations an inspector adopts during an inspection of the manufacture of sterile medicinal products. It reflects the most important changes introduced in the revised Annex 1, but is not intended to address all changes in the revision.

  - Document history
  - Purpose and scope
  - Basics
  - Definitions and abbreviations
  - New texts and their interpretation
  - Revision history

  - **Guide to Good Manufacturing Practice for Medicinal Products – Annexes**
  - Annex 1 - Manufacture of sterile medicinal products
  - Annex 2 - Manufacture of biological medicinal products for human use
  - Annex 3 - Manufacture of radiopharmaceuticals
  - Annex 4 - Manufacture of veterinary medicinal products other than immunologicals
  - Annex 5 - Manufacture of immunological veterinary medical products
  - Annex 6 - Manufacture of medicinal gases
  - Annex 7 - Manufacture of herbal medicinal products
  - Annex 8 - Sampling of starting and packaging materials
  - Annex 9 - Manufacture of liquids, creams and ointments
  - Annex 10 - Manufacture of pressurised metered dose aerosol preparations for inhalation
  - Annex 11 - Computerised systems
  - Annex 12 - Use of ionising radiation in the manufacture of medicinal products
  - Annex 13 - Manufacture of investigational medicinal products
  - Annex 14 - Manufacture of products derived from human blood or human plasma
  - Annex 15 - Qualification and validation
  - Annex 16 - Qualified person and batch release
  - Annex 17 - Parametric release
  - Annex 18 - GMP guide for active pharmaceutical ingredients (This Annex no longer exists)
  - Annex 19 - Reference and retention samples
  - Annex 20 - Quality risk management
  - Glossary

EudraLex - Volume 4 - Good Manufacturing Practice (GMP) guidelines

Volume 4 of "The rules governing medicinal products in the European Union" contains guidance for the interpretation of the principles and guidelines of good manufacturing practices for medicinal products for human and veterinary use laid down in Commission Directives 91/356/EEC, as amended by Directive 2003/94/EC, and 91/412/EEC respectively.

Eudralex V4 is made up of the following parts:
➢ Introduction
➢ Part I - Basic Requirements for Medicinal Products
➢ Part II - Basic Requirements for Active Substances used as Starting Materials
➢ Part III - GMP related documents

Introduction
The Commission Directive 2003/94/EC, of 8 October 2003, set out the principles and guidelines of good manufacturing practice in respect of medicinal products for human use and investigational medicinal products for human use.

Part I - Basic Requirements for Medicinal Products

Chapter 1 - Pharmaceutical Quality System
Chapter 2 - Personnel
Chapter 3 - Premise And Equipment
Chapter 4 - Documentation
Chapter 5 - Production
Chapter 6 - Quality Control
Chapter 7 - Outsourced Activities
Chapter 8 - Complaints And Product Recall
Chapter 9 - Self Inspection

Part II - Basic Requirements for Active Substances used as Starting Materials

Basic requirements for active substances used as starting materials

Part III - GMP related documents

Site Master File
Q9 Quality Risk Management
Q10 Note for Guidance on Pharmaceutical Quality System
MRA Batch Certificate

Annexes
Annex 1-Manufacture of Sterile Medicinal Products
Annex 2- Manufacture of Biological active substances and Medicinal Products for Human
Annex 3- Manufacture of Radiopharmaceuticals

Annex 4- Manufacture of Veterinary Medicinal Products other than Immunological Veterinary Medicinal Products

Annex 5- Manufacture of Immunological Veterinary Medicinal Products

Anne 6- Manufacture of Medicinal Gases

Annex 7- Manufacture of Herbal Medicinal Products

Annex 8-Sampling of Starting and Packaging Materials

Annex 9-Manufacture of Liquids, Creams and Ointments

Annex 10-Manufacture of Pressurised Metered Dose Aerosol Preparations for Inhalation

Annex 11-Computerised Systems

Annex 12-Use of Ionising Radiation in the Manufacture of Medicinal Products

Annex 13-Manufacture of Investigational Medicinal Products

Annex 14-Manufacture of Products derived from Human Blood or Human Plasma

Annex 15-Qualification and validation (into operation since 1 October 2015)

Annex 16-Certification by a Qualified Person and Batch Release

Annex 17-Parametric Release

Annex 19-Reference and Retention Samples

FDA

FDA publishes regulations and guidance documents for industry in the Federal Register. FDA's website also contains links to the CGMP regulations and guidance documents various resources to help drug companies comply with the law. FDA also conducts onsite audits and public outreach through presentations at national and international meetings and conferences on the subject of CGMP requirements.

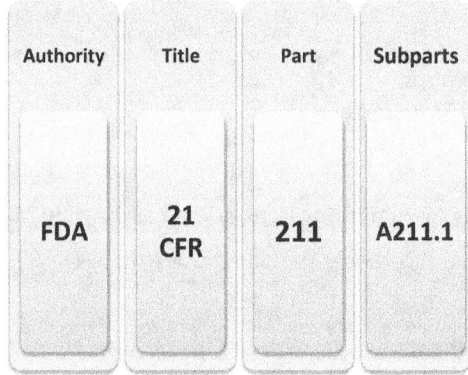

Figure 2: The FDA organises its regulations under titles. Within titles there parts and subparts.

Pharmaceutical quality affects every American. FDA regulates the quality of pharmaceuticals very carefully. The main regulatory standard for ensuring pharmaceutical quality is the Current Good Manufacturing Practice (CGMPs) regulation for human pharmaceuticals. Consumers expect that each batch of medicines they take will meet quality standards so that they will be safe and effective. Most people, however, are not aware of CGMPs, or how FDA assures that drug manufacturing processes meet these basic objectives. Recently, FDA has announced a number of regulatory actions taken against drug manufacturers based on the lack of CGMPs. This paper discusses some facts that may be helpful in understanding how CGMPs establish the foundation for drug product quality.

PART 211   Current Good Manufacturing Practice For Finished Pharmaceuticals

§ 211.115 - Reprocessing.

Subpart G--Packaging and Labeling Control
§ 211.122 - Materials examination and usage criteria.
§ 211.125 - Labeling issuance.
§ 211.130 - Packaging and labeling operations.
§ 211.132 - Tamper-evident packaging requirements for over-the-counter (OTC) human drug products.
§ 211.134 - Drug product inspection.
§ 211.137 - Expiration dating.

Subpart H--Holding and Distribution
§ 211.142 - Warehousing procedures.
§ 211.150 - Distribution procedures.

Subpart I--Laboratory Controls
§ 211.160 - General requirements.
§ 211.165 - Testing and release for distribution.
§ 211.166 - Stability testing.
§ 211.167 - Special testing requirements.
§ 211.170 - Reserve samples.
§ 211.173 - Laboratory animals.
§ 211.176 - Penicillin contamination.

Subpart J--Records and Reports
§ 211.180 - General requirements.
§ 211.182 - Equipment cleaning and use log.
§ 211.184 - Component, drug product container, closure, and labeling records.
§ 211.186 - Master production and control records.
§ 211.188 - Batch production and control records.
§ 211.192 - Production record review.
§ 211.194 - Laboratory records.
§ 211.196 - Distribution records.
§ 211.198 - Complaint files.

Subpart K--Returned and Salvaged Drug Products
§ 211.204 - Returned drug products.
§ 211.208 - Drug product salvaging.

World Health Organisation GMP Guideline Annexes
The WHO Essential Medicines and Health Products (EMP) Department works with countries to promote affordable access to quality, safe and effective medicines, vaccines, diagnostics and other medical devices. As part of this effort, the WHO publishes a number of guidance annexes that describe best practice quality requirements for specific areas within the life science industry.

List of WHO GMP annexes:

- WHO good manufacturing practices for pharmaceutical products: main principles
  Annex 2, WHO Technical Report Series 986, 2014
- Active pharmaceutical ingredients (bulk drug substances)
  Annex 2, WHO Technical Report Series 957, 2010
- Active pharmaceutical ingredients - bulk drug substances: Additional clarifications and explanations
- Pharmaceutical excipients
  Annex 5, WHO Technical Report Series 885, 1999
- WHO good manufacturing practices for sterile pharmaceutical products
  Annex 6, WHO Technical Report Series 961, 2011
- WHO good manufacturing practices for biological products
  Annex 3, WHO Technical Report Series 996, 2016
- WHO good manufacturing practices for blood establishments (jointly with the Expert Committee on Biological Standardization)
  Annex 4, WHO Technical Report Series 961, 2011
- Pharmaceutical products containing hazardous substances
  Annex 3 WHO Technical Report Series 957, 2010
- Investigational pharmaceutical products for clinical trials in humans
  Annex 7, WHO Technical Report Series 863, 1996
- Herbal medicinal products
  Annex 3, WHO Technical Report Series 937, 2006
- Radiopharmaceutical products
  Annex 3, WHO Technical Report Series 908, 2003
- Water for pharmaceutical use
  Annex 2, WHO Technical Report Series 970, 2012
- WHO guidelines on good manufacturing practices for heating, ventilation and air-conditioning systems for non-sterile pharmaceutical dosage forms
  Annex 5, WHO Technical Report Series 961, 2011
- Validation
  Annex 4, WHO Technical Report Series 937, 2006
- Guidelines on good manufacturing practices: validation, Appendix 7: non-sterile process validation
  Annex 3, WHO Technical Report Series 992, 2015

International Council for Harmonisation, ICH, GMP Guide
The International Council for Harmonization of (Technical Requirements) for Pharmaceuticals for Human Use (ICH) brings together the regulatory authorities and pharmaceutical industry to discuss scientific and technical aspects of drug registration. Since its inception in 1990, ICH has gradually evolved, to respond to the increasingly global face of drug development.
- ICH Q7 Good Manufacturing Practice Guide for Active Pharmaceutical Ingredients

1.8 Documentation and Records

Good documentation is an essential part of the quality assurance system and is relevant across all departments and functions within a manufacturing company. Controlled documents define the specifications and procedures for all materials and methods of manufacture and control strategies.

General Requirements

➢ Documents should be designed, prepared, reviewed, approved and distributed in accordance with approved processes.
➢ Documents should comply with relevant parts of the manufacturing and marketing authorizations.
➢ Documents should be approved, signed and dated by the appropriate responsible persons. No document should be changed without authorization and approval.
➢ Documents should have unambiguous contents: the title, nature and purpose should be clearly stated. They should be laid out in an orderly fashion and be easy to check.
➢ Documents must not allow any error to be introduced through the reproduction process.

Good Documentation Practices

This section provides an easy-to-understand guide to the subject of Good Documentation Practices. Good Documentation Practices (commonly abbreviated to GDP or GDocP) is a term used to describe standards by which documents are created, modified and maintained. The need for GDP is driven by the general requirement of GMP (Good Manufacturing Practices)

GDP is a practical skill that is required within the life science sector (medical device, pharmaceutical and so on). It can be broadly divided into two streams; GDocP practices and how they apply to electronic document and secondly, GDocP for handwritten entries including initial and dating and recording of data and test results by hand. GDocP is fundamental in achieving compliance to Good Manufacturing Practices (GMP). It is required in the U.S. by the FDA's Code of Federal Regulations and in Europe by the governing body EudraLex. If GDocP is not practiced it jeopardises the integrity of data and written records. This can lead to the falsification of data which is a serious regulatory offence. Admittedly, implementing and maintaining GDP takes time, effort and resources, however, there are some benefits that come with it. Most importantly, Good Documentation Practice is an expected practice in regulated industry as trust and ethics are fundamental to business.

It is important to maintain accurate, complete, up-to-date and consistent information to ensure patient safety and reduce any potential risk to patients. Practicing GDP equally helps to reduce observations raised on inadequate documentation practices at times of audit by regulated bodies such as the FDA. It helps to improve communication and efficiency within companies. If GDP is not followed it can call into question other processes and procedures within a company.

Documentation Creation

The principles of GDP should be applied at the document creation stage. As most people are familiar with softcopy or electronic documents, some of these points are obvious but nonetheless need to be made. All documents should be electronically written and not handwritten except for execution of protocols, test results and adding entries. Documents that are approved controlled should be:

Accurate and free from errors
Have revision or version controlled
Should have an effective date or date of release

Approval of Documents

Document approval must be completed by trained and appropriately experienced personnel. Often companies will use an approval matrix which explains which people are required to approve each document. For example, an EHS (Environment Health and Safety) officer would be required to approve a risk assessment.

Signatures

A signature on any document is legally binding so remember to read and understand what is being signed for. Every signature should also include the date in the correct format. If a signature appears within the same document alongside initials, substituting a full signature with initials and date is generally acceptable. This practice is common when large documents are being completed.

Date and Time Format

A standardised approach to dates and times is important especially within large global organisations. For instance, in the USA, the norm is to place the month before the date, whereas in Ireland and Great Britain it is common to write the day of the month followed by the month. Most companies would define their date and time format in an SOP or procedure.

The date and time format can also be configured in Word documents and Excel worksheets to align with a companies preferred date and time format.

Handwritten Entries

When a handwritten entry is required such as a signature or a test result, indelible ink must be used. Many companies will have an SOP or procedure that states the specific ink colour required. If an entry of a test result or test data isn't completed at the time of execution, this constitutes a late entry. Backdating an entry or signature is forbidden. Always use the correct and current date.

How Are Mistakes Corrected?

This is a critical area of GDP. Failure to follow the requirements of GDP when correcting mistakes is the most common failure when it comes to documentation in industry. The method of correcting mistakes using GDP allows for a person looking at the document for the first time to clearly see the original entry and the corrected entry. This maintains the integrity of the document. In order to identify the changes and corrections, certain rules must be followed. No overwriting is allowed and white-out or Tipp-Ex is not allowed. The correct way to make any changes or corrections by hand is shown in the diagram below.

Accuracy

Accuracy of information provided in documents is critical in the life science industry. As the end user is a patient, inaccurate records or documents could cause serious injury or death. Controlled documents are also legal documents and could be called upon if recalls, litigation or investigations arise.

Many documents used in the manufacture of medical devices are designed to record information or test results. These test results are then used to disposition (pass or fail) batches of product. Inaccurate information could risk the release and distribution of defective product. This has a potential impact on both the business and the patient or user.

Blank Spaces or Blank Fields

On completion of a document such as a logbook or record, no blanks spaces should be left unfilled. This is to avoid late entries and also to prevent confusion. Blank spaces or blank fields should have a diagonal line drawn neatly across the space, the letters "N/A" written and the entry signed and dated. If the reason for "N/A" is not evident then it is wise to include an explanatory note or sentence.

Data Transcription

Transcribing is the process of transferring data from one source to another. This is often required when raw data is involved. When data is in raw format it may need to be entered into a Microsoft Excel sheet. When transcribing data remember that all original raw data must be stored in case it is needed at a future date. After the data is transcribed it must be verified by a second person to check for any errors or omissions.

Revision Control

Controlled documents should always have a version number or revision number electronically on each page of the document. This is similar to books which always list what edition they are e.g. first edition or second edition. Revision control is a key element of the Quality Management Systems in place in regulated industries. As the need for changes in the document arises, the controlled document can be amended/updated. With each update the version number revises also. Some companies will use alphabetic revision control and to a lesser extent numeric revision control (Version A, Version B or Version 01, Version 02).

Management of Attachments

Attachments to controlled documents can include training records, data sheets, lab results and so on. It is important that attachments are identified for traceability purposes. If the attached becomes detached from the main document, the attachment should be identifiable. It is best practice to include a reference number on the attachment if available. If the attachment consists of several pages, each page should be numbered in Page X of Y format if not electronically done so. And remember, hand written entries must be accompanied by a signature and date. Always use staples to attach documents together. Glue or paper clips are not acceptable.

Management of Documents through Their Lifecycle

GDP applies to all the different stages of a document's lifecycle. These stages include creation, review, approval, issuing, completion of records, revision, updating, retirement and storage. Storage a.k.a. retention is an important stage and often a legal requirement for medical devices and pharmaceutical products. For consumer OTC medicines a 5-year retention of quality records often suffices. For implants such as TKRs or Total Knee Replacements, a 90-year retention period is required. This ensures that traceability and a quality record is available if the need arises.

Test Results

This section provides an overview on the correct handling of test results. Test results can be generated from various types of product testing such as visual inspection, dimensional inspection and chemical analysis. The recording of all test results should be completed on an approved form. This is to ensure that the correct information is being recorded and the same approach is taken by different people who might have to complete testing.

Calculations

There are different ways calculations can be completed. Many simple calculations can be done by an individual using a calculator, alternatively, a software package such as Minitab or an Excel sheet can be used to complete complex calculations. It should be clear to the reader what calculation is required, what the formula is and how the calculation is completed.

If the formula used is not included on the sheet, it should be referenced in a controlled document. Care is also required when recording numbers of several decimal places in length, as rounding error can be introduced.

Units of Measurement

The most important thing to remember is consistency in units of measurement when recording data or making calculations. Consult your company procedure if available to determine the correct units of measurement. Many U.S. companies use imperial units e.g. inches, pounds etc. In Europe the International System of Units or SI is used, e.g. millimetres and kilograms.

Batch Records

Batch records document critical information relating to the manufacture of products. Depending on the product, it can include dispensed weights of raw materials. It may also include critical parameters, times and dates of critical steps, in process test results and so on.

- ➢ Batch records should be reviewed and checked for:
- ➢ Accuracy
- ➢ Legibility
- ➢ Correct document version
- ➢ Completeness
- ➢ Correct references to supporting documents
- ➢ a unique batch or identification number
- ➢ be dated
- ➢ signed when issued/approved

With reference to ICH Q7, the following requirements are specified in a clear and concise format beneficial to the manufacturer.

"Documentation of completion of each significant step in the batch production records (batch production and control records) should include: – Dates and, when appropriate, times;

– Identity of major equipment (e.g., reactors, driers, mills, etc.) used;

– Specific identification of each batch, including weights, measures, and batch numbers of raw materials, intermediates, or any reprocessed materials used during manufacturing; – Actual results recorded for critical process parameters;

– Any sampling performed;

– Signatures of the persons performing and directly supervising or checking each critical step in the operation;

– In-process and laboratory test results;
– Actual yield at appropriate phases or times;
– Description of packaging and label for intermediate or API;
– Representative label of API or intermediate if made commercially available;
– Any deviation noted, its evaluation, investigation conducted (if appropriate) or reference to that investigation if stored separately;

– Results of release testing."

Laboratory Records

Laboratory control records should include complete data derived from all tests conducted to ensure compliance with established specifications and standards, including examinations and assays. A description of samples received for testing, including the material name or source, batch number or other distinctive code, date sample was taken, and, where appropriate, the quantity and date the sample was received for testing; ICH Q7 states the following requirements.

– A statement of or reference to each test method used;
– A statement of the weight or measure of sample used for each test as described by the method; data on or cross-reference to the preparation and testing of reference standards, reagents and standard solutions;
– A complete record of all raw data generated during each test, in addition to graphs, charts, and spectra from laboratory instrumentation, properly identified to show the specific material and batch tested;
– A record of all calculations performed in connection with the test, including, for example, units of measure, conversion factors, and equivalency factors;
– A statement of the test results and how they compare with established acceptance criteria; –

The signature of the person who performed each test and the date(s) the tests were performed; and

– The date and signature of a second person showing that the original records have been reviewed for accuracy, completeness, and compliance with established standards.

Complete records should also be maintained for:

– Any modifications to an established analytical method;
– Periodic calibration of laboratory instruments, apparatus, gauges, and recording devices;

– All stability testing performed on APIs; and
– Out-of-specification (OOS) investigations.
(Ref: ICH, Q7.)

2.0 PERSONNEL

Introduction

Personnel are central to the application of CGMP and compliance to regulations. A every level throughout an organisation, people interact with materials, equipment and processes in order to deliver products to the market and patient. Personnel must therefore be suitably qualified and equipped to carry out their responsibilities effectively.

Provisions in guidance and regulations are therefore made for Personnel in a Quality Management System. Despite advances in automation and computerised systems, people are centrally involved in day to day decisions. For this reason there must be sufficient and suitability qualified personnel to carry out all the tasks. Individual responsibilities should be clearly defined and understood by the persons concerned and recorded formally in procedures and job descriptions. It may be an obvious point; however, manufacturers must ensure an adequate number of personnel with the necessary qualifications and practical experience are resourced to manufacturing. Having a broad base of people with the experience, knowledge and skills reduces the risk of quality issues. Responsibilities placed on any one individual should not be so extensive as to present any risk to quality. Personnel should have specific duties recorded in written descriptions and adequate authority to carry out its responsibilities. Its duties may be delegated to designated deputies with a satisfactory level of qualifications.

Personal Hygiene

All personnel should be trained in the practices of personal hygiene. A high level of personal hygiene should be observed by all those concerned with manufacturing processes. Personnel should be instructed to wash their hands before entering production areas. Signage should be in place along with hand washing facilities. Hand washing demonstrations and training should be provided by a suitably qualified QC analyst or Microbiologist. Any person experiencing an illness or exhibiting open lesions or wounds that may adversely affect the quality of products should not be allowed to handle starting materials, packaging materials, in-process materials or medicines until the condition is no longer a risk to quality or patient safety. Direct contact should be avoided between the operator's hands and starting materials, primary packaging materials and intermediate or bulk product.

Contamination Control

The philosophy of containment control requires it to be applied across all inputs that make up a facility- equipment, processes, and utilities and so on. Containment is primarily concerned with keeping things in- preventing product or processing agents from egressing into the surrounding atmosphere. Ensuring adequate containment protects personnel who interact with the process, equipment and systems. Aseptic processing often deals with biological agents or compounds that may be harmful to operators or technicians. A secondary concern of containment is protection of the environment. Containment also compliments efforts in contamination prevention. As with Aseptic processing the risk to the patient and product must be at the forefront of activity. Risk based approaches and tools should be used to identify potential risks and put in place adequate controls and mitigations. Any assessment should take into account all the following systems:

- Facility layout
- Drainage Systems
- HVAC requirements
- Location and adequacy of utilities
- Personnel flow and procedures for entering and leaving
- Behavioral requirements of personnel in the clean room
- Flow of materials and products to prevent cross-contamination and mix-ups between products and between dirty and clean or sterile and non-sterile equipment and products
- Design to avoid cross-contamination when manufacturing live biological agents, e.g. local exhaust air HEPA filtration, dedicated air handling units.

Material Transfer

Material transfer from the outside of cleanrooms to the inside is completed via material air locks or hatches. Material air locks and hatches ensure that there is clear separation between controlled clean areas and less clean areas. Many suppliers provide products that are double bagged. This provides an added level of control when transferring materials. The outer bag can be removed within the air lock thus providing a clean inner product. Material air locks also allow the sanitization of products. Tools and other items must be clean and dirt free.

Controls that prevent personnel from the clean area and less clean area been present in the material air lock at the same time. This can be achieved by training and educating staff on the importance of contamination control. A simple visual check of the air lock to confirm it is vacant can be done in order to avoid mixing of personal from different zones. Decontamination procedures are necessary to ensure materials or tools entering the controlled area are decontaminated.

Material Air lock considerations:
➢ Interlocked doors
➢ Access control
➢ Sanitation/ Cleaning procedure
➢ Double or triple bagged products
➢ Dedicated trolley for air locks

Disinfection and Cleaning Agents

When materials are been transferred via an air lock, consideration must be given to the status of materials and products. As a rule, no cardboard or unnecessary paper should enter a cleanroom. Wooden pallets are not acceptable as they can carry dirt and microorganisms and wood cannot be sanitized due to its porous nature. Soft fabric cases often used to carry tools should also be avoided as the material can carry dirt and grease. Cleaning and disinfecting agents should be tested and approved prior to their use onsite. The choice of agents should be backed up with studies that demonstrate the effectiveness of disinfectants and cleaning agents.

Gown up Areas

Gowning rooms are designed in order to minimize contamination and facilitate the orderly change over from street clothes to scrubs and/or gowns. Hand washing facilities help reduce the risk of humans carrying unwanted microorganisms into the aseptic processing area. The design of the room should result in clear separation between the less clean side and the clean side. This can be achieved with a step over segregating the two areas.

Other features of gowning rooms should include:

➢ Storage lockers for street clothes
➢ Gown and Garment storage
➢ Body Length mirrors
➢ Hand Washing /Drying and disinfection facilities

3.0 MATERIALS MANAGEMENT

3.1 Introduction

The key theme of effective materials management is control of materials from incoming stage through the manufacturing process. Specifications and testing support the control of materials ensuring they are meeting the key quality requirements to allow consistent manufacturing and quality products.

Starting Materials

The designated name of the product and the internal code reference where applicable;

> - Manufacturers batch number
> - the status of the contents (e.g. quarantined, on test, released)
> - the expiry date or a date beyond which retesting is necessary

Packaging Materials

The purchase, handling and control of primary and printed packaging materials should be as for starting materials. Particular attention should be paid to printed packaging materials. They should be stored in secure conditions so as to exclude the possibility of unauthorized access. Roll feed labels should be used wherever possible. Cut labels and other loose printed materials should be stored and transported in separate closed containers so as to avoid mix ups. Packaging materials should be issued for use only by designated personnel following an approved and documented procedure.

Intermediate

Intermediate products can be simply described as raw materials that may have been mixed and processed to some degree or other. Intermediate and bulk products should be kept under appropriate conditions and must be used within specified dates and according to specifications.

Finished Product

Finished products should be held in quarantine until their final release, after which they should be stored as usable stock under conditions established by the manufacturer. For the approval and maintenance of suppliers of active substances and excipients, the following is required:

Active substances

Supply chain traceability should be established and the associated risks, from active substance starting materials to the finished medicinal product, should be formally assessed and periodically verified. Appropriate measures should be put in place to reduce risks to the quality of the active substance.

The supply chain and traceability records for each active substance (including active substance starting materials) should be available and be retained by the EEA based manufacturer or importer of the medicinal product. Audits should be carried out at the manufacturers and distributors of active substances to confirm that they comply with the relevant good manufacturing practice and good distribution practice requirements. The holder of the manufacturing authorisation shall verify such compliance either by himself or through an entity acting on his behalf under a contract.

Further audits should be undertaken at intervals defined by the quality risk management process to ensure the maintenance of standards and continued use of the approved supply chain.

Excipients

Excipients and excipient suppliers should be controlled appropriately based on the results of a formalised quality risk assessment in accordance with the European Commission 'Guidelines on the formalised risk assessment for ascertaining the appropriate Good Manufacturing Practice for excipients of medicinal products for human use'.

For each delivery of starting material the containers should be checked for integrity of package, including tamper evident seal where relevant, and for correspondence between the delivery note, the purchase order, the supplier's labels and approved manufacturer and supplier information maintained by the medicinal product manufacturer. The receiving checks on each delivery should be documented.

Prevention of Cross-contamination

Cross-contamination should be prevented by attention to design of the premises and equipment. This should be supported by attention to process design and implementation of any relevant technical or organizational measures, including effective and reproducible cleaning processes to control risk of cross-contamination.

A Quality Risk Management process, which includes a potency and toxicological evaluation, should be used to assess and control the cross-contamination risks presented by the products manufactured. Factors including; facility/equipment design and use, personnel and material flow, microbiological controls, physicochemical characteristics of the active substance, process characteristics, cleaning processes and analytical capabilities relative to the relevant limits established from the evaluation of the products should also be taken into account. The outcome of the Quality Risk Management process should be the basis for determining the necessity for and extent to which premises and equipment should be dedicated to a particular product or product family. This may include dedicating specific product contact parts or dedication of the entire manufacturing facility.

Suggested Technical Measures
- ➢ Dedicated manufacturing facility (premises and equipment)

- Self-contained production areas having separate processing equipment and separate heating, ventilation and air-conditioning (HVAC) systems. It may also be desirable to isolate certain utilities from those used in other areas
- Design of manufacturing process, premises and equipment to minimize opportunities for cross-contamination during processing, maintenance and cleaning
- Use of "closed systems" for processing and material/product transfer between equipment
- Use of physical barrier systems, including isolators, as containment measures
- Controlled removal of dust close to source of the contaminant e.g. through localised extraction
- Dedication of equipment, dedication of product contact parts or dedication of selected parts which are harder to clean (e.g. filters), dedication of maintenance tools;
- Use of single use disposable technologies
- Use of equipment designed for ease of cleaning
- Appropriate use of air-locks and pressure cascade to confine potential airborne contaminant within a specified area
- Minimising the risk of contamination caused by recirculation or re-entry of untreated or insufficiently treated air
- Use of automatic clean in place systems of validated effectiveness
- For common general wash areas, separation of equipment washing, drying and storage areas.

Suggested Organisational Measures
- Dedicating the whole manufacturing facility or a self-contained production area on a campaign basis (dedicated by separation in time) followed by a cleaning process of validated effectiveness
- Keeping specific protective clothing inside areas where products with high risk of cross-contamination are processed
- Cleaning verification after each product campaign should be considered as a detectability tool to support effectiveness of the Quality Risk Management approach for products deemed to present higher risk
- Depending on the contamination risk, verification of cleaning of non product contact surfaces and monitoring of air within the manufacturing area and/or adjoining areas in order to demonstrate effectiveness of control measures against airborne contamination or contamination by mechanical transfer
- Specific measures for waste handling, contaminated rinsing water and soiled gowning;
- Recording of spills, accidental events or deviations from procedures
- Design of cleaning processes for premises and equipment such that the cleaning processes in themselves do not present a cross-contamination risk
- Design of detailed records for cleaning processes to assure completion of cleaning in accordance with approved procedures and use of cleaning status labels on equipment and manufacturing areas
- Use of common general wash areas on a campaign basis

3.2 Rejection and Re-use of Materials

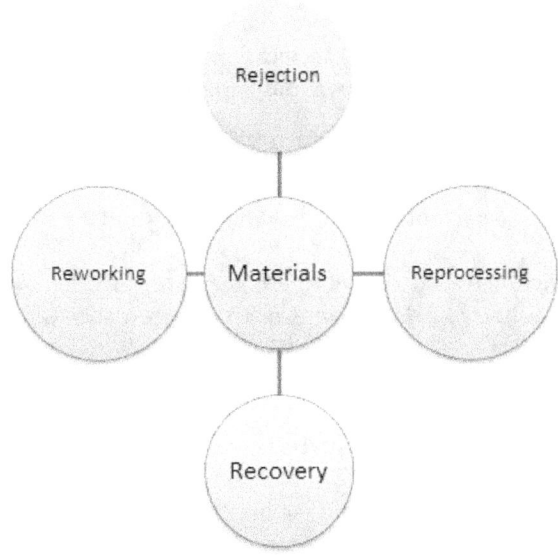

Rejection

Intermediates and components failing to meet established specifications should be identified as such and quarantined according to a procedure. These items can be reprocessed or reworked as described below.

Reprocessing

Reprocessing by repeating a manufacturing step or a chemical or physical process of an established manufacturing process is generally considered acceptable.

Reprocessing should involve evaluation to ensure that the quality of product and must not adversely impact the safety of the finished product.

Recovery of Materials and Solvents

Recovery (e.g. from mother liquor or filtrates) of reactants, intermediates, or the API is considered acceptable, provided that approved procedures exist for the recovery and the recovered materials meet specifications suitable for their intended use.

Returns

Records of returned intermediates or APIs should be maintained. For each return, documentation should include:

- Name and address of the consignee
- Intermediate or API, batch number, and quantity returned
- Reason for return
- Use or disposal of the returned intermediate or API

Testing Of Materials

The tests performed should be recorded adequately. EU GMP V4 Part 1 Chapter 6: Quality Control recommends the following information as a minimum.

- ➢ Name of the material or product and, where applicable, dosage form
- ➢ Batch number and, where appropriate, the manufacturer and/or supplier
- ➢ References to the relevant specifications and testing procedures
- ➢ Test results, including observations and calculations, and reference to any certificates of analysis
- ➢ Dates of testing
- ➢ Initials of the persons who performed the testing
- ➢ Initials of the persons who verified the testing and the calculations, where appropriate

Sampling Checklist

The sample taking should be done and recorded in accordance with approved written procedures that describe:

- ➢ The method of sampling
- ➢ The equipment to be used
- ➢ . The amount of the sample to be taken
- ➢ Instructions for any required sub-division of the sample
- ➢ The type and condition of the sample container to be used
- ➢ The identification of containers sampled
- ➢ Any special precautions to be observed, especially with regard to the sampling of sterile
- ➢ or noxious materials
- ➢ The storage conditions
- ➢ Instructions for the cleaning and storage of sampling equipment

4.0 VALIDATION

4.1 Introduction

The term 'Validation Lifecycle' refers to the entire lifecycle, beginning with the initial requirements of a product or process and identifying CPPs and CQAs. The cycle continues through Commissioning and Qualification (C&Q), PQ and PV, requalification and ending with decommissioning or the end of life of a product line. Process Validation is defined as "establishing documented evidence which provides a high degree of assurance that a specific process consistently produces a product meeting its predetermined specifications and quality attributes."

PICS/s

The PIC/s guidance on Validation is limited to describing the basic requirements of a validation program such as having a validation procedure, qualifying a new process so that it is "suitable and "consistent".

PE 009-13 (Part I), Chapter 5, states:

➢ *"Validation studies should reinforce Good Manufacturing Practice and be conducted in accordance with defined procedures. Results and conclusions should be recorded.*
➢ *When any new manufacturing formula or method of preparation is adopted, steps should be taken to demonstrate its suitability for routine processing. The defined process, using the materials and equipment specified, should be shown to yield a product consistently of the required quality.*
➢ *Significant amendments to the manufacturing process, including any change in equipment or materials, which may affect product quality and/or the reproducibility of the process should be validated.*
➢ *Processes and procedures should undergo periodic critical revalidation to ensure that they remain capable of achieving the intended results."*

Eudralex

"Validation studies should reinforce Good Manufacturing Practice and be conducted in accordance with defined procedures. Results and conclusions should be recorded.

When any new manufacturing formula or method of preparation is adopted, steps should be taken to demonstrate its suitability for routine processing. The defined process, using the materials and equipment specified, should be shown to yield a product consistently of the required quality.

Significant amendments to the manufacturing process, including any change in equipment or materials, which may affect product quality and/or the reproducibility of the process, should be validated.

Processes and procedures should undergo periodic critical re-validation to ensure that they remain capable of achieving the intended results." (Ref: GMP V4 Part I, Chapter 5)

FDA

For Medical Devices, FDA 21 CFR Subpart G, Part 820 specifies the requirements for full verification or validation:

(*a*) *Where the results of a process cannot be fully verified by subsequent inspection and test, the process shall be validated with a high degree of assurance and approved according to established procedures. The validation activities and results, including the date and signature of the individual(s) approving the validation and where appropriate the major equipment validated, shall be documented.*

(*b*) *Each manufacturer shall establish and maintain procedures for monitoring and control of process parameters for validated processes to ensure that the specified requirements continue to be met.*

(*1*) *Each manufacturer shall ensure that validated processes are performed by qualified individual(s).*
(*2*) *For validated processes, the monitoring and control methods and data, the date performed, and, where appropriate, the individual(s) performing the process or the major equipment used shall be documented.*

(*c*) *When changes or process deviations occur, the manufacturer shall review and evaluate the process and perform revalidation where appropriate. These activities shall be documented.*

For medicinal/ pharmaceutical drug products 21 CFR 211.100(a) and 211.110(a) requires that drug products be produced with a high degree of assurance of meeting all the attributes they are intended to possess.

WHO
WHO GMP guidance requires that, each pharmaceutical company identifies what qualification and validation work is required to prove that the critical aspects of their particular operation are controlled. A validation plan should identify and describe what activities are required to be undertaken.

Annex 2 identifies the key requirements with regard to qualification and validation to ensure:

➤ the premises, supporting utilities, equipment and processes have been designed in accordance with the requirements for GMP (design qualification or DQ)

➤ the premises, supporting utilities and equipment have been built and installed in compliance with their design specifications (installation qualification or IQ)

➤ the premises, supporting utilities and equipment operate in accordance with their design specifications (operational qualification or OQ)

➤ a specific process will consistently produce a product meeting its predetermined specifications and quality attributes (process validation or PV, also called performance qualification or PQ).

ICH
ICH Q7 provides arguably provides the greatest amount of detail with regard to validation in a GMP environment. Similar to other organisations, ICH requires a company to develop a "Validation Policy" to describe and document approaches to validations etc. ICH gives good guidance on validation in respect of Active pharmaceutical ingredients and the importance of critical process parameters and critical quality attributes. The approaches to Process Validation (Prospective, Concurrent etc.) also align with FDA requirements.

Key considerations for Validation of APIs
> Defining critical product attributes of APIs
> Identifying process parameters that could affect the critical quality attributes of API's
> Process Validation (PV should provide documented evidence that the process, operated within established parameters, can perform effectively and reproducibly to produce an intermediate or API meeting its predetermined specifications and quality attributes.

ICH Q7 key definitions regarding Qualification / Validation

"Design Qualification (DQ): documented verification that the proposed design of the facilities, equipment, or systems is suitable for the intended purpose.

Installation Qualification (IQ): documented verification that the equipment or systems, as installed or modified, comply with the approved design, the manufacturer's recommendations and/or user requirements.

Operational Qualification (OQ): documented verification that the equipment or systems, as installed or modified, perform as intended throughout the anticipated operating ranges.

Performance Qualification (PQ): documented verification that the equipment and ancillary systems, as connected together, can perform effectively and reproducibly based on the approved process method and specifications."

The Four Types of Process Validation
Process validation is a regulatory requirement of Good Manufacturing Practices (GMPs) for both pharmaceuticals (21CFR 211) and medical devices (21 CFR 820).

Prospective validation
Establishing documented evidence in advance of process implementation that a process or system operates as intended. This is the preferred approach and is most common when new products must be validated before commercial manufacturing.

Concurrent validation
Establishing documented evidence that a processes operates as intended, based on information generated during process implementation. Concurrent means that the outputs are performance of the system is monitored at the same time a manufacturing which can include commercial lots.

Retrospective validation
Retrospective validation is used for facilities or processes that have not completed formal Validation. Historical data or a retrospective review can provide the evidence that the process or facility is operated as intended. This type of validation is uncommon.

Revalidation
Revalidation involves the re-execution of validation activities in order to maintain a validated state. This can be a result of substantial changes to Product attributes or specification or changes to the manufacturing process itself. Other reasons a partial or full revalidation may be required involve instances where product quality issues have increased.

Stages of Process Validation

Process validation can be divided into in three stages:

Stage 1 – Process Design: The commercial manufacturing process is defined during this stage based on knowledge gained through development and scale-up activities.

Stage 2 – Process Qualification: During this stage, the process design is evaluated to determine if the process is capable of reproducible commercial manufacturing.

Stage 3 – Continued Process Verification: Ongoing assurance is gained during routine production that the process remains in a state of control.

Before any batch from the process is commercially distributed for use by consumers, a manufacturer should have gained a high degree of assurance in the performance of the manufacturing process such that it will consistently produce APIs and drug products meeting those attributes relating to identity, strength, quality, purity, and potency. The assurance should be obtained from objective information and data from laboratory-, pilot-, and/or commercial scale studies. Information and data should demonstrate that the commercial manufacturing process is capable of consistently producing acceptable quality products within commercial manufacturing conditions. A successful validation program depends upon information and knowledge from product and process development. This knowledge and understanding is the basis for establishing an approach to control of the manufacturing process that result in products with the desired quality attributes. Understanding variation and knowing how to detect and control it is therefore a key element to maintaining robust processes and systems.

Stage 1 — Process Design: Process design is the activity of defining the commercial manufacturing process that will be reflected in planned master production and control records. The goal of this stage is to design a process suitable for routine commercial manufacturing that can consistently deliver a product that meets its quality attributes. Building and Capturing Process Knowledge and Understanding Generally, early process design experiments do not need to be performed under the CGMP conditions required for drugs intended for commercial manufacturing and supply.

Stage 2 (process qualification) and Stage 3 (continued process verification). They should, however, be conducted in accordance with sound scientific methods and principles, including good documentation practices. Decisions and justification of the controls should be sufficiently documented and internally reviewed to verify and preserve their value for use or adaptation later in the lifecycle of the process and product. Although often performed at small-scale laboratories, most viral inactivation and impurity clearance studies cannot be considered early process design experiments. Viral and impurity clearance studies intended to evaluate and estimate product quality at commercial scale should have a level of quality unit oversight that will ensure that the studies follow sound scientific methods and principles and the conclusions are supported by the data.

Product development activities provide key inputs to the process design stage, such as the intended dosage form, the quality attributes, and a general manufacturing pathway. Process information available from product development activities can be leveraged in the process design stage. The functionality and limitations of commercial manufacturing equipment should be considered in the process design, as well as predicted contributions to variability posed by different component lots, production operators, environmental conditions, and measurement systems in the production setting. Design of Experiment (DOE) studies can help develop process knowledge by revealing relationships, including multivariate interactions, between the variable inputs (e.g., component characteristics or process parameters) and the resulting outputs (e.g., in-process material, intermediates, or the final product).

Risk analysis tools can be used to screen potential variables for DOE studies to minimize the total number of experiments conducted while maximizing knowledge gained. These activities also provide information that can be used to model or simulate the commercial process. Computer-based or virtual simulations of certain unit operations or dynamics can provide process understanding and help avoid problems at commercial scale. It is important to understand the degree to which models represent the commercial process, including any differences that might exist, as this may have an impact on the relevance of information derived from the models. It is essential that activities and studies resulting in process understanding be documented. Establishing a Strategy for Process Control Process knowledge and understanding is the basis for establishing an approach to process control for each unit operation and the process overall.

Strategies for process control can be designed to reduce input variation, adjust for input variation during manufacturing (and so reduce its impact on the output), or combine both approaches. Process controls address variability to assure quality of the product. Controls can consist of material analysis and equipment monitoring at significant processing. Decisions regarding the type and extent of process controls can be aided by earlier risk assessments, then enhanced and improved as process experience is gained. The planned commercial production and control records, which contain the operational limits and overall strategy for process control, should be carried forward to the next stage for confirmation.

Stage 2 — Process Qualification During the process qualification (PQ) stage of process validation, the process design is evaluated to determine if it is capable of reproducible commercial manufacture.
This stage has two elements: (1) design of the facility and qualification of the equipment and utilities and (2) process performance qualification (PPQ). During Stage 2, CGMP-compliant procedures must be followed. Successful completion of Stage 2 is necessary before commercial distribution. Products manufactured during this stage, if acceptable, can be released for distribution.

Design of a Facility and Qualification of Utilities and Equipment Proper design of a manufacturing facility is required under part 211, subpart C, of the CGMP regulations on Buildings and Facilities. It is essential that activities performed to assure proper facility design and commissioning precede PPQ.

Here, the term qualification refers to activities undertaken to demonstrate that utilities and equipment are suitable for their intended use and perform properly. These activities necessarily precede manufacturing products at the commercial scale. Qualification of utilities and equipment generally includes the following activities:

➢ Selecting utilities and equipment construction materials, operating principles, and performance characteristics based on whether they are appropriate for their specific uses.
➢ Verifying that utility systems and equipment are built and installed in compliance with the design specifications (e.g., built as designed with proper materials, capacity, and functions, and properly connected and calibrated).
➢ Verifying that utility systems and equipment operate in accordance with the process requirements in all anticipated operating ranges. This should include challenging the equipment or system functions while under load comparable to that expected during normal operation.

It should also include the performance of interventions, stoppage, and start-up as is expected during routine production. Operating ranges should be shown capable of being held as long as would be necessary during routine production. Qualification of utilities and equipment can be covered under individual plans or as part of an overall project plan.

The plan should consider the requirements of use and can incorporate risk management to prioritize certain activities and to identify a level of effort in both the performance and documentation of qualification activities.

Design of facilities and the qualification of utilities and equipment, personnel training and qualification, and verification of material sources (components and container/closures), if not previously accomplished.

Review and approval of the protocol by appropriate departments and the quality unit.. PPQ Protocol Execution and Report Execution of the PPQ protocol should not begin until the protocol has been reviewed and approved by all appropriate departments, including the quality unit. Any departures from the protocol must be made according to established procedure or provisions in the protocol. Such departures must be justified and approved by all appropriate departments and the quality unit before implementation (§ 211.100).

The commercial manufacturing process and routine procedures must be followed during PPQ protocol execution (§§ 211.100(b) and 211.110(a)). The PPQ lots should be manufactured under normal conditions by the personnel routinely expected to perform each step of each unit operation in the process. Normal operating conditions should include the utility systems (e.g., air handling and water purification), material, personnel, environment, and manufacturing procedures. A report documenting and assessing adherence to the written PPQ protocol should be prepared in a timely manner after the completion of the protocol and execution of the PPQ activity.

This report should:
➢ Discuss and cross-reference all aspects of the protocol.
➢ Summarize data collected and analyze the data, as specified by the protocol.
➢ Evaluate any unexpected observations and additional data not specified in the protocol.

- Summarize and discuss all manufacturing non-conformances such as deviations, aberrant test results, or other information that has bearing on the validity of the process.
- Describe in sufficient detail any corrective actions or changes that should be made to existing procedures and controls.
- State a clear conclusion as to whether the data indicates the process met the conditions established in the protocol and whether the process is considered to be in a state of control. If not, the report should state what should be accomplished before such a conclusion can be reached. This conclusion should be based on a documented justification for the approval of the process, and release of lots produced by it to the market in consideration of the entire compilation of knowledge and information gained from the design stage through the process qualification stage.
- Include all appropriate department and quality unit review and approvals.

Stage 3 — Continued Process Verification

The goal of the third validation stage is continual assurance that the process remains in a state of control (the validated state) during commercial manufacture. A system or systems for detecting unplanned departures from the process as designed is essential to accomplish this goal. Adherence to the CGMP requirements, specifically, the collection and evaluation of information and data about the performance of the process, will allow detection of undesired process variability. Evaluating the performance of the process identifies problems and determines whether action must be taken to correct, anticipate, and prevent problems so that the process remains in control (§ 211.180(e)). An ongoing program to collect and analyze product and process data that relate to product quality must be established (§ 211.180(e)).

The data collected should include relevant process trends and quality of incoming materials or components, in-process material, and finished products. The data should be statistically trended and reviewed by trained personnel. The information collected should verify that the quality attributes are being appropriately controlled throughout the process. We recommend that a statistician or person with adequate training in statistical process control techniques develop the data collection plan and statistical methods and procedures used in measuring and evaluating process stability and process capability.

Procedures should describe some references that may be useful include the following:

- ASTM E2281-03 "Standard Practice for Process and Measurement Capability Indices,"

- ASTM E2500-07 "Standard Guide for Specification, Design, and Verification of Pharmaceutical and Biopharmaceutical Manufacturing Systems and Equipment,"

- ASTM E2709-09 "Standard Practice for Demonstrating Capability to Comply with a Lot Acceptance Procedure."

Production data should be collected to evaluate process stability and capability. The quality unit should review this information. If properly carried out, these efforts can identify variability in the process and/or signal potential process improvements. Good process design and development should anticipate significant sources of variability and establish appropriate detection, control, and/or mitigation strategies, as well as appropriate alert and action limits. However, a process is likely to encounter sources of variation that were not previously detected or to which the process was not previously exposed. Many tools and techniques, some statistical and others more qualitative, can be used to detect variation, characterize it, and determine the root cause.

Leading manufacturers should use quantitative, statistical methods whenever appropriate and feasible. Scrutiny of intra-batch as well as inter-batch variation is part of a comprehensive continued process verification program under § 211.180(e).

Best practices ensures continued monitoring and sampling of process parameters and quality attributes at the level established during the process qualification stage until sufficient data are available to generate significant variability estimates. These estimates can provide the basis for establishing levels and frequency of routine sampling and monitoring for the particular product and process.

Monitoring can then be adjusted to a statistically appropriate and representative level. Process variability should be periodically assessed and monitoring adjusted accordingly. Variation can also be detected by the timely assessment of defect complaints, out-of specification findings, process deviation reports, process yield variations, batch records, incoming raw material records, and adverse event reports.

Production line operators and quality unit staff should be encouraged to provide feedback on process performance. We recommend that the quality unit meet periodically with production staff to evaluate data, discuss possible trends or undesirable process variation, and coordinate any correction or follow-up actions by production.

Data gathered during this stage might suggest ways to improve and/or optimize the process by altering some aspect of the process or product, such as the operating conditions (ranges and set-points), process controls, component, or in-process material characteristics. A description of the planned change, a well-justified rationale for the change, an implementation plan, and quality unit approval before implementation must be documented (§ 211.100). Depending on how the proposed change might affect product quality, additional process design and process qualification activities could be warranted. Maintenance of the facility, utilities, and equipment is another important aspect of ensuring that a process remains in control. Once established, qualification status must be maintained through routine monitoring, maintenance, and calibration procedures and schedules (21 CFR part 211, certain manufacturing changes may call for formal notification to the Agency before implementation, as directed by existing regulations (see, e.g., 21 CFR 314.70 and 601.12). The equipment and facility qualification data should be assessed periodically to determine whether re-qualification should be performed and the extent of that re-qualification. Maintenance and calibration frequency should be adjusted based on feedback from these activities.

4.2 Equipment Validation

Validation is a legal and regulatory requirement for the manufacture of medicinal products and medical devices. The area of Validation can be sub-divided into two elements. Equipment Qualification (EQ) and Process Validation. Equipment qualification ensure that equipment operates as intended and is installed in accordance with the manufacturers recommendation. Process Validation involves the provision of documented evidence to confirm a particular process performs consistency and meets pre-determined specifications.

All equipment that can impact the quality of product is subject to Validation, hence equipment and systems used in aseptic manufacturing must undergo equipment and process validation. Installation Qualification (IQ) protocols should cover verification that all utilities are installed correctly to the manufacturers recommendations. All sitting and mechanical connections should also be confirmed as adequate. Other key tests and verifications includes:

- ➤ Documentation of Materials of Construction (MOC)
- ➤ Calibration of equipment based instrumentation
- ➤ Spare parts listing
- ➤ Preventative maintenance schedule creation
- ➤ Electrical installation verification
- ➤ Health and Safety assessment
- ➤ Ergonomic Assessment
- ➤ Documenting Software and Hardware
- ➤ Backup of software
- ➤ Backup of Recipes (Sterilization, Bio-decontamination etc.)

The system User Requirements Specification (URS) should provide the basis of testing and must be fulfilled during the course of Validation.

The ultimate goal of Equipment Qualification is to ensure that equipment is fit for its intended use. Therefore, equipment is validated to confirm it functions as intended and meets all requirements to manufacture product safely and consistently. FDA requires that "Each manufacturer shall ensure that all equipment used in the manufacturing process meets specified requirements and is appropriately designed, constructed, placed and installed to facilitate maintenance, adjustment, cleaning and use". In other words all manufacturing equipment, support facilities, measuring and test equipment shall be "qualified". (FDA 21 CFR 820.70 (g))

Equipment Qualification Protocols are developed to document this testing and hence provide evidence on the functionality and consistency of the equipment. There are two distinct parts within the scope of Equipment Qualification, Installation Qualification and Operational Qualification. Often these subparts are abbreviated to IQ and OQ. Other combinations such as IOQE and IQ/OQ can be encountered within industry. This is often defined in a company's procedure or SOP relating to Equipment Validation.

A User Requirements Specification (URS) is often used to document the "specified requirements" of a particular piece of equipment. A URS can then be used as an input document when equipment qualification is required. While a URS document can be extensive covering areas such as equipment functionality, utility requirements, safety features, software specs etc. not all requirements documented in a URS will need to be verified or validated. Critical requirements should be identified early and should always be verified.

In short Equipment Qualification is confirmation via documented evidence that the particular requirements for a specific intended use can be consistently fulfilled under anticipated conditions.

5.0 COMPLAINTS AND RECALLS

5.1 Introduction

Patients and medical professionals are encouraged to be vigilant and report any potentially defective or suspect products. Manufacturers must have a system in place to receive complaints and a written procedure that details how complaints are reviewed and acted upon. Manufacturers must also have a system that facilitates the recall of products known or suspected to be on the market.

Suspect or defective product

Complaint received

Assessment of complaint according to written procedure

Decision making

Recall (if required)

General points:

> ➢ Record all details of complaint

> ➢ Document the investigation

> ➢ Document all decisions

> ➢ Review type and amount of complaints regularly

> ➢ Communicate with Competent authorities as appropriate

Investigation

Q7 Guidance states the following with regard to investigations:

> ➢ *The information reported in relation to possible quality defects should be recorded, including all the original details. The validity and extent of all reported quality defects should be documented and assessed in accordance with Quality Risk Management principles in order to support decisions regarding the degree of investigation and action taken.*

> If a quality defect is discovered or suspected in a batch, consideration should be given to checking other batches and in some cases other products, in order to determine whether they are also affected. In particular, other batches which may contain portions of the defective batch or defective components should be investigated.

> Quality defect investigations should include a review of previous quality defect reports or any other relevant information for any indication of specific or recurring problems requiring attention and possibly further regulatory action.

> The decisions that are made during and following quality defect investigations should reflect the level of risk that is presented by the quality defect as well as the seriousness of any non-compliance with respect to the requirements of the marketing authorisation/product specification file or GMP. Such decisions should be timely to ensure that patient and animal safety is maintained, in a way that is commensurate with the level of risk that is presented by those issues.

> As comprehensive information on the nature and extent of the quality defect may not always be available at the early stages of an investigation, the decision-making processes should still ensure that appropriate risk-reducing actions are taken at an appropriate time-point during such investigations. All the decisions and measures taken as a result of a quality defect should be documented.

> Where human error is suspected or identified as the cause of a quality defect, this should be formally justified and care should be exercised so as to ensure that process, procedural or system-based errors or problems are not overlooked, if present.

> Appropriate CAPAs should be identified and taken in response to a quality defect. The effectiveness of such actions should be monitored and assessed.

> Quality defect records should be reviewed and trend analyses should be performed regularly for any indication of specific or recurring problems requiring attention.

5.2 Recalls

Recalls should be managed and co-ordinated by a responsible person with adequate support from a wider team to handle all aspects of the recall or complaint. This responsible person typically is independent of the sales and marketing organisation.

6.0   RISK MANAGEMENT

6.1 Introduction

This chapter provides guidance on the application and interpretation of ISO 14971, Risk management system. The standard provides a framework on how to apply risk management to medical devices. However, its principles and core definitions can be applied across MedTech, pharmaceuticals, biopharma and combinations products. Risk management principles and techniques are frequently applied to assess process risks, design risks and product risks. Hence, risk management is visible at across the many stages of GxP, GMP and life sciences. There are regulatory requirements specific to products and intended use that require risk management in addition to the benefits to provide to the manufacturer and the protection to the patient or end user.

The standard sets out the requirements on development, implementation and maintenance of a risk management process for medical devices. It is acknowledged as the principal standard to use when conducting medical device risk management activities. Risk is the combination of the probability of occurrence of harm and the severity of that harm Source – ISO/IEC Guide 51:1999).The term "risk" within the scope of the ISO 14971 International Standard on refers to safety or performance requirements of the medical device or meeting applicable regulatory requirements.

The various risks presented by a particular device depends substantially on its intended purpose and the effectiveness of the risk management techniques used in the design, manufacture and subsequent use by the end user. Principles of risk management are best applied using a Process and Iterative Approach. A process works to ensure requirements are documents, instructions and templates are in place and roles and responsibilities are defined. An effective risk management process will often have many work instructions or SOPs providing the requirements for aspects of risk management such as PFMEAs, risk planning, risk review, post marketing surveillance and so on.

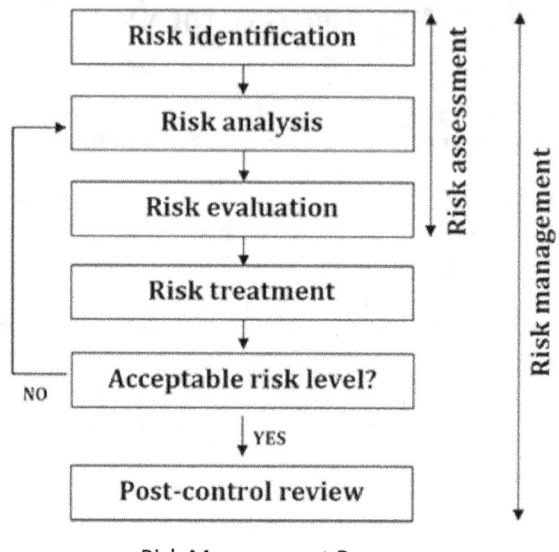

Risk Management Process

Harm
Harm occurs when ac physical injury or damage to the health of people, or damage to property or the environment.

Hazard
A hazard is a potential source of harm.

Hazardous situation
circumstance in which people, property or the environment are exposed to one or more harms.

Risk
combination of the probability of occurrence of harm and the severity of that harm

Risk Analysis
The systematic use of available information to identify hazards and to estimate the risk. Risk analysis also refers to the analysis of the various sequences of events that can produce hazardous situations and harm

Risk Control
process in which decisions are made and measures implemented by which risks are reduced

Risk Estimation
A process used to assign values to the probability of occurrence of harm and the severity of that harm

Risk Management

systematic application of management policies, procedures and practices to the tasks of analysing, evaluating, controlling and monitoring risk

Safety
freedom from unacceptable risk

Severity
the measure of the possible consequences of a hazard

General Requirements
General requirements for risk management system are covered in Clause 4 under sub-clause 4.1 Risk management process. The manufacturer shall establish, implement, document and maintain an ongoing process for:

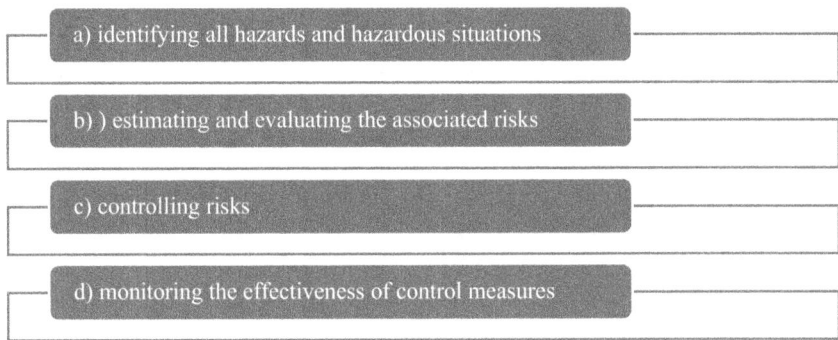

a) identifying all hazards and hazardous situations

b)) estimating and evaluating the associated risks

c) controlling risks

d) monitoring the effectiveness of control measures

Requirements of the manufacturer

In addition, the following ground rules apply:
-The Risk management applies throughout the life cycle of the medical device.
-The process shall include the following elements:
        -RISK ANALYSIS
        -RISK EVALUATION
        -RISK CONTROL
        -PRODUCTION AND POST-PRODUCTION ACTIVITIES
These areas are discussed in further detail further on in this book.

Regulations and Standards

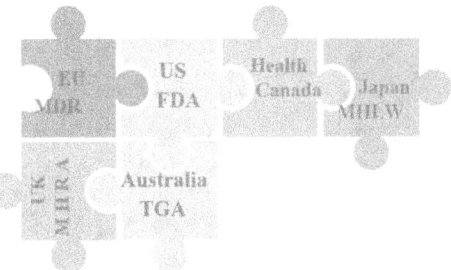

Key Competent Authorities

Manufacturers must have a Risk Management process defined in approved procedures along with documentation for the medical device products demonstrating the risk process is applied during development and over the lifecycle also.

Regulatory bodies worldwide recommend the application of ISO 14971 Medical devices — Application of Risk Management to Medical Devices. In addition, other medical device industry standards require risk management to achieve compliance. One such examples is Annex B of ISO 10993-1:2018 which provides guidance on the risk management approach for identification of biological hazards and so on. These requirements (of international standards) can be considered risk control measures. International standards are considered to be generally acknowledged state of the art. With this in mind, for Risk Management the manufacturer can make the task of identifying residual risk that bit easier by applying the standard and working through the particular requirements. Above all, the role of risk management is to provide medical devices that are safe and meet the intended use. In addition, a manufacturer may apply product standards or process standards that require specific requirements to be met and demonstrated. If a manufacturer applies ISO 14971, Risk Management and other product and process related standards relevant to the medical device, the residual risks related to the safety hazards and hazardous situation can be considered acceptable unless there is objective evidence to the contrary (e.g. adverse events, complaints, recalls)

Risk Analysis

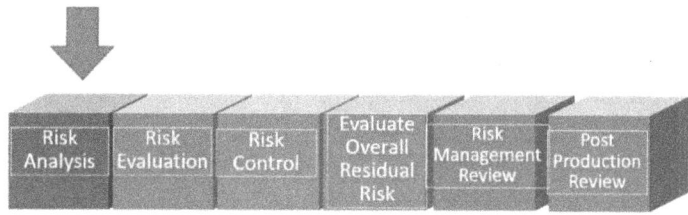

Risk analysis process can be sub divided into 4 parts which are detailed in ISO 14971. These include:

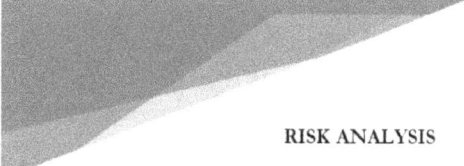

RISK ANALYSIS

(1) description of the intended-use of the medical device and reasonably foreseeable misuses
(2) Identification of the characteristics of the medical device that are related to safety
(3) Identification of hazards and hazardous situations associated with the medical device
(4) estimation of risk for each hazardous situation

Understanding the Intended use of a medical device is fundamental as it determines the proper application and use of the device. Designers aim to properly define the intended use as it then allows them to focus on what specific requirements will deliver such a device, meeting the user requirements and intended use. Intended is concerned with (1) medical indication, (2) patient population, (3) user profile (e.g. doctor or lay person) (4) part of the body or tissue the device is concerned with and (5) the use environment.

Reasonably foreseeable misuse

This is a scenario when a medical device is used in a way that it was not intended or designed to be used as set out by the manufacturer. Situations of reasonably foreseeable misuse are understood as situations that can anticipated based on human behavior. (hence reasonably foreseeable).

As part of risk management reasonably foreseeable misuses should be identified by the manufacturer. These can be identified in a number of ways which include:

- During product realization and Design and Development)
- Simulated studies such as Usability Engineering Studies
- Customer Complaints or adverse events during Post market monitoring.

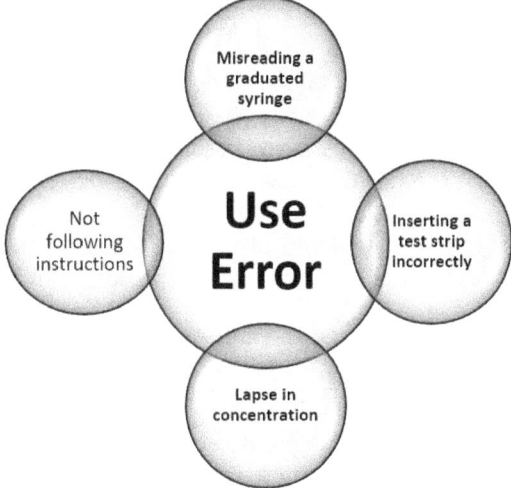

Identification of characteristics related to safety

Certain characteristics of a medical device can affect safety! It is up to the manufacturer to identify the performance requirements or the functions of the medical devices that fulfil the intended use or if hazardous situations can occur that may impact the performance of safety. Refer to section on Risk Analysis for information on questions used to Identify hazards and characteristics related to safety

Identification of hazards and hazardous situations

| Hazardous situations resulting from faults |
| Hazardous situations resulting from random faults |
| Hazardous situations resulting from systematic faults |
| Hazardous situations arising from security |

A hazard = potential source of a harm

Hazards can be identified from both the intended use and also any reasonably foreseeable misuse. As previously mentioned, Annex A of ISO 14971:2019 provides details on the characteristics relating to safety. In turn, these characteristics can help in identifying hazards and hazardous situations. Note: Annex C of ISO 14971:2019 provides guidance that can help in identifying hazards and sequences of events that can lead to hazardous situations.

Hazardous Situations

Hazardous situations resulting from faults
If a hazardous situation occurs due to a fault condition, the probability of a fault occurring is not the same as the probability of the occurrence of harm. A fault condition may initiate a sequence of events, however this may not necessarily, result in a hazardous situation. Therefore, a hazardous situation does not result in harm always.

Hazardous situations resulting from random faults
Random faults can be a result physical or chemical corrosion, contamination and mechanical wear-out.

Hazardous situations resulting from systematic faults
The term systematic fault intends to describe when a series of actions/environmental conditions or inputs combine to cause a fault condition. It can be caused by an error in any activity but normally remains latent unless the combination of conditions lead to the fault happening.

Identification of hazards and hazardous situations
This section summaries the questions used in the identification of hazardous situations. (Blue Graphic). For each question, a practical insight is provided to assist in the application of the questions. A common practice of manufacturers is to list the questions and responses in risk documented such as a risk analysis document or a design risk analysis.

Risk Analysis Techniques:

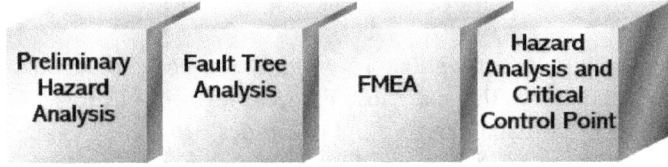

Preliminary Hazard Analysis (PHA) Risk Assessment technique favored particularly early-on in the development stage to identify hazards and hazardous situations & events that can cause harm when few of the details of the medical device design are known.

Fault Tree Analysis (FTA) utilized in the development stage or development process. It is useful in identification of hazards/hazardous situations but also the ranking of them. It also records risk control measures and the potential affects on the end user

Failure Mode and Effects Analysis, (FMEA) is where the effects or consequences of individual process steps (PFMEA) or components (DFMEA) are systematically identified. Each Failure mode has an associated effect or consequence. The cause of the failure mode is documented. The severity, occurrence and detection are also recorded along current available controls. Additional risk controls can thus reduce the overall risk. RPN.

Hazard Analysis and Critical Control Point (HACCP) also involves the identification, evaluation, and control of risks. As each hazard is listed, critical control points and critical limits are assigned to each hazard. This is then followed by monitoring, corrective actions, verification and record keeping.

Preliminary Hazard Analysis (PHA)

Preliminary Hazard Analysis, PHA is methodology with the objective of identifying (1)hazards, (2) hazardous situations and events that can cause harm for a given activity, facility or system. PHA requires the recording of potential hazards and hazardous situations by review of the process, equipment or system. When the hazards and hazardous situations are listed probabilities that they lead to harm should be calculated or estimated. The identification of possible risk control measures is also an important part of completing the analysis.

Fault Tree Analysis (FTA)

Fault Tree Analysis (FTA) can be used to analyze hazards identified. Fault tree analysis (FTA) is referred to as a top-down, deductive failure analysis methodology where an undesired state of a system (or product or process) is analyzed using Boolean logic to present a series of lower-level events. In deductive reasoning, a conclusion is reached by applying logic rules to reach a conclusion. FTA is utilized across many industries such as aerospace, nuclear and reliability engineering applications.

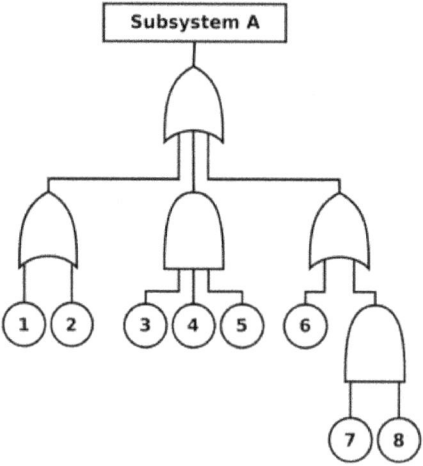

Failure Mode and Effects Analysis (FMEA)

The FMEA methodology looks consequences of an individual failure modes that are identified and evaluated based on the process steps (PFMEA) or the component level (DFMEA). FMEAs looking at the manufacturing process or assembly are known as Process FMEAs. If it related to the Design or in which way the device could fail, it is deemed a Design FMEA. If the FMEA is focused on the use or foreseeable misuse of the device it is referred to as a Use FMEA.

Failure Mode

- Potential failures or non-conformances in a product or process. The way in which a process, system, or component could potentially fail.

Cause

- The action (lack of action of mechanism that creates the failure mode condition

Effect

- the consequences which a failure mode can have on the process/product or user

Hazard Analysis and Critical Control Point (HACCP)

HACCP is methodology that works to identify hazards and hazardous situations and in turn control the risk by monitoring the critical control points identified in a the manufacturing process. To recap on the meanings of hazards and hazardous situations. A hazard can be defined as a potential to cause harm. A hazardous situation occurs when person or user is exposed to one or more hazards. It usually requires a sequence of events for a hazardous situation to occur. Such as a defective sterile barrier damaged and the patient going on to use the product (essentially not following the instructions for use).

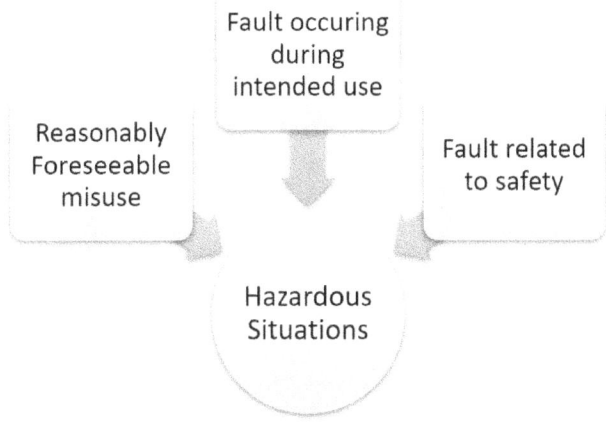

HACCP can be applied using the below steps:

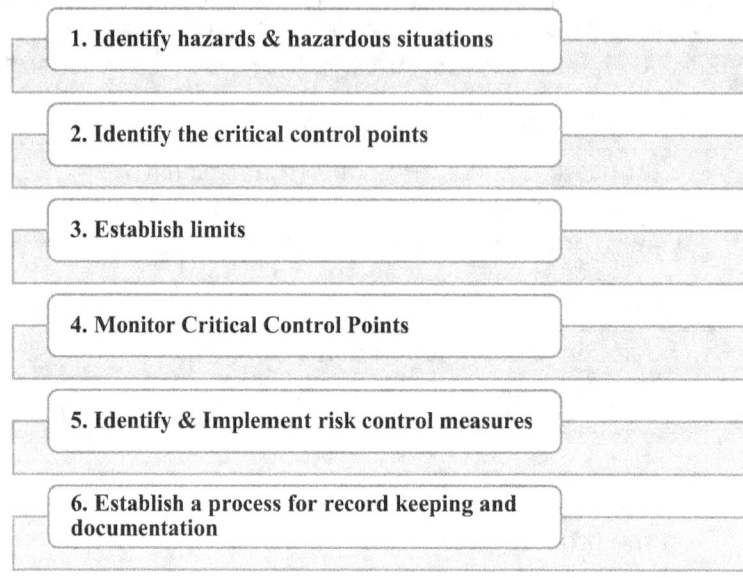

1. **Identify hazards & hazardous situations**

2. **Identify the critical control points**

3. **Establish limits**

4. **Monitor Critical Control Points**

5. **Identify & Implement risk control measures**

6. **Establish a process for record keeping and documentation**

1. Identify hazards and hazardous situations - For any manufacturing process, the first step is to identify at which points in the process potential hazards exist.
2. Identify critical control points- these are points in which specific controls can be applied to ensure risk is reduced to acceptable levels or so that the risk is limited.

3. Establish limits -Each process step will have limits that are based on quality requirements. For example, there may be a dimensional limit (minimum diameter, maximum diameter).
4. Monitor CCPs-monitoring the control limits will tell you if the process is within the critical limits and if there is a risk to product. Monitoring may be continuous via automation or it may be more infrequent such as manual inspection or manual audit.
5. Corrective Actions- identify and track corrective actions to closure and review effectiveness
6. Establish documentation and record keeping procedures

In conclusion, when applying the HACCP methodology, attention on the continuing control and monitoring of the identified hazards and hazardous situations is paramount.

Risk Estimation / Evaluation

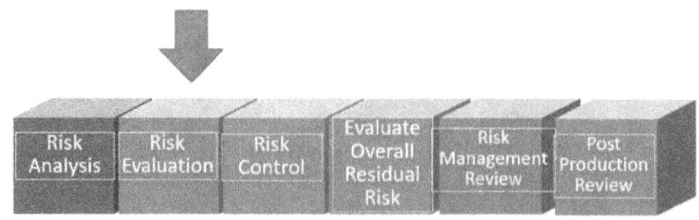

Probability
Qualitative estimation of probability is where descriptions are used to estimate probability, High (often), medium (sometimes) low (rare, unlikely). A Quantitative approach is where information or data is used to estimate probability, for example, parts per million. The quantitative approach, due to the fact it is normally based on data can provide more thorough risk estimation and evaluation.

Risk estimation
Risk estimation involves the analysis of the probability of occurrence of a harm and the severity of the harm.

"For each identified hazardous situation, the manufacturer shall estimate the associated risk (s) using available information/data."

Risk estimation incorporates an analysis of (1) probability of occurrence of harm and the (2)severity of the harm.

The qualitative and quantitative systems used for categorization of probability of occurrence of harm and severity of harm need to be recorded in the risk management file to comply with ISO 14971.

High- likely to happen, often or frequently, always

Medium- can happen, but not frequently

Low -unlikely to happen, rare and remote

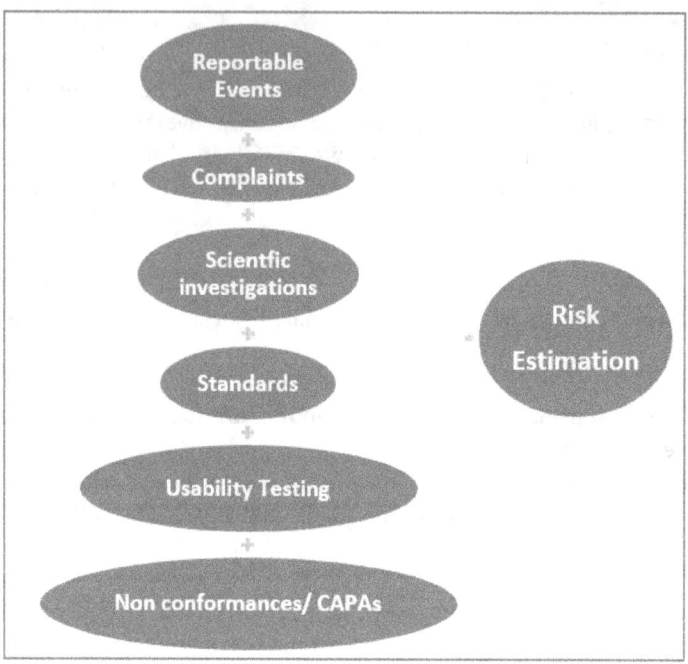

Sources of data for estimation of risk

Risk Control

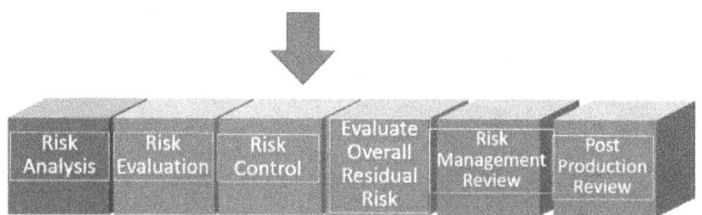

If risk is too high then further risk control measures must be implemented to reduce the risk to an acceptable level. A number of actions can be taken in order to further reduce risk including: (1) changing the design to reduce risk- safety that is built-in or inherent in the design is very effective. (2) introducing protective measures in the device or the manufacturing process, (3) including a warning statement into the instructions for use (IFU). (4) Use of symbols on labelling and cartoning- information on safety if communicated and used by the user can provide mitigation also.

Risk Acceptability

ISO 14971 requires companies to document a Policy and develop criteria for risk acceptability. The policy provides instruction on how to establish the criteria for acceptability of the overall residual risk. It should address individual residual risks and the risk-benefit ratio or analysis also.
ISO 14971 requires a policy for establishing the criteria for risk acceptability. The policy can include (1) purpose, (2) scope, (3) factors and considerations for determining acceptable risk, (4) approaches to risk control and (5) requirements for approval and review.

- The purpose should detail the specific goals of the policy for establishing criteria for risk acceptance/acceptability.
- Factors and considerations for risk acceptability- Applicable regulatory requirements and international standards for the medical device,

Criteria for risk acceptability

Criteria for risk acceptability should be established in advance of any risk management activity or execution of the risk management plan so that guidance is available in determining acceptable risk. The policy is normally is included in a risk management procedures or other quality document.

Evaluation of overall residual risk and acceptability

An evaluation of the overall residual risk and acceptability must be completed in accordance with the risk policy. ISO 14971 specifies that both the method and the criteria be stated in the risk management plan.

Clause 8:

"Evaluation of overall residual risk After all risk control measures have been implemented and verified, the manufacturer shall evaluate the overall residual risk posed by the medical device, taking into account the contributions of all residual risks, in relation to the benefits of the intended use, using the method and the criteria for acceptability of the overall residual risk defined in the risk management plan. If the overall residual risk is judged acceptable, the manufacturer shall inform users of significant residual risks and shall include the necessary information in the accompanying documentation in order to disclose those residual risks.

If the overall residual risk is not judged acceptable in relation to the benefits of the intended use, the manufacturer may consider implementing additional risk control measures or modifying the medical device or its intended use. Otherwise, the overall residual risk remains unacceptable. The results of the evaluation of the overall residual risk shall be recorded in the risk management file. Compliance is checked by inspection of the risk management file and the accompanying documentation"

Role of Management

Top Management are responsible for establishing and maintaining an effective risk management process. The term Top Management should draw the level of commitment to the handling of Risk and the risk management process. Top management can assign personnel, make resources available and set the priorities for the business. By making Top management responsible for the risk management process an effective process that is maintained is supported.

Risk Management Plan

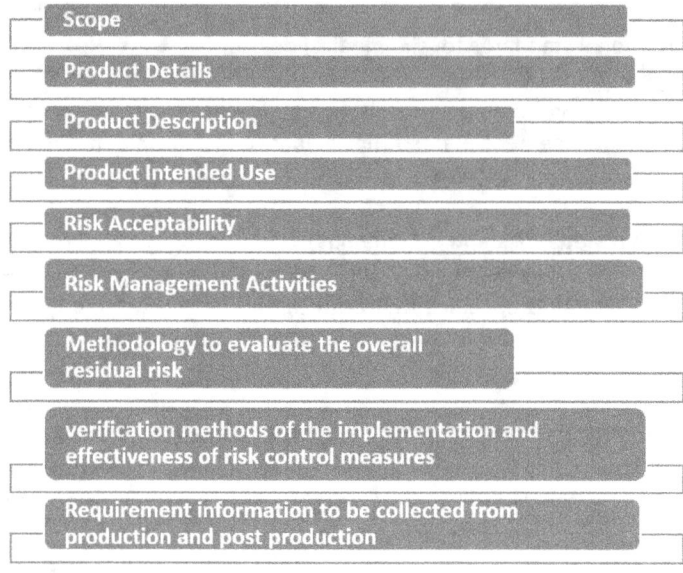

Scope:

Product Details
Product name(s) (While it may seem obvious, some brands or products may have many similar but different products on the market. For example device may have varying levels of complexity and functionality pitched a different prices on the market, e.g. Product *Pro, Product *Gold, Product* extra.

Product SKU
The Risk management plan should set out the product or products included in the scope of the plan. Sometimes one plan will cover a family of products or subgroup or products. This practice is common and acceptable. Defining the scope of the risk management plan is a requirement of ISO 14971.

Product description
A product description should be informative and consistent between marketing materials, patient literature and design documentation.

Intended use
The intended use of the product must be included in the Risk management plan.

Roles and Responsibilities
Roles and responsibilities and required to be included per ISO 14971. Defining such roles and responsibilities ensures the right expertise and invested in risk management and helps to ensure successful application of the methods and verifications that ensure an effective risk management process.

The plan must identify all of the Risk Management activities planned during the lifecycle of the product. These include:

Risk Management Plan inputs
- Management Review
- Design Management
- Risk Management
- Change Management
- Complaint Handling
- CAPA Management
- Clinical Evaluation Reports
- Adverse reportable events

The risk management plan details by what means (how) and when (new products, product design changes, periodic review) the risk management activities will be reviewed for a medical device or defined family. Key sections within a compliant risk management plan include the method of review, the responsible individuals/ functions, how the output of the review is managed. Results are then reflected in the risk management report.

Criteria for risk acceptability
The ISO 14971 Risk management process requires a manufacturer's to have a policy that requires the establishment of what is deemed acceptable risk- also known as risk acceptability. To ensure this policy is established with the right parameters and for it to remain impartial, the criteria for risk acceptance should be created before commencing the risk assessment.

Method to evaluate overall residual risk and criteria for acceptability
 The overall residual risk and the criteria for its acceptability are based on the manufacturer's policy for establishing criteria for risk acceptability. Per ISO 14971, the method and the criteria must be stated in the risk management plan for the particular medical device.

Verifications methods and activities
The risk management plan should specify how the verification activities are executed, or alternatively it should reference another document that provides the details. The plan should specify the methods that are used to verify that any risk Control measures are implemented and to what degree they are effective. In addition, the overall residual risk must have a method of evaluation and criteria for the acceptability of the overall residual risk.

Post production and Post Marketing Requirements
Post-production information becomes an input into Risk Management activities for the product and also features in the risk management plan. The type of post marketing surveille (sources of information, analysis of information) should be appropriate for the product covered in the plan.

Risk management Review and Reporting
The risk management review is an important step before the commercial release of the medical device and required per ISO 13485, clause 9, The final results of the risk management process, as obtained by executing the risk management plan, are reviewed. The risk management report contains the results of this review and is a crucial part of the risk management file.

| the risk management plan has been implemented |
| the overall residual risk is acceptable |
| collect and review information of production and postproduction |
| results are recorded in the risk management report |

The report serves as the high-level document that provides evidence that the manufacturer has ensured that the risk management plan has been satisfactorily fulfilled and the results confirm that the required objective has been achieved. Subsequent reviews of the execution of the risk management plan and updates of the risk management report can be needed during the life cycle of the medical device, as a result of the execution of production and post-production activities.

When

- The risk management review should be completed when the implementation of the plan is completed along with verification of the risk control measures. This should occur in advance of product release of commercial product.

What

- The risk management report is the concluding output of all risk management activities.

Schedule

- The risk management file must be maintained over the life cycle of the product. The risk mangament report should therefore be reviewed at appropriate intervals to ensure it is accurate and current. Other scenarios that require review or update of the risk management report would be a change in the maufacturing process that would require re-validation of the process or a new product type been added to a current range

Effectiveness

- The risk management process must be suitable, fit for purpose and demonstrated as effective.

Severity

Severity is based on the harm that could result from the use of a particular medical device. Severity must be included and documented in the Risk Management File in accordance with the requirements of ISO 14971:2019.

Risk Management File

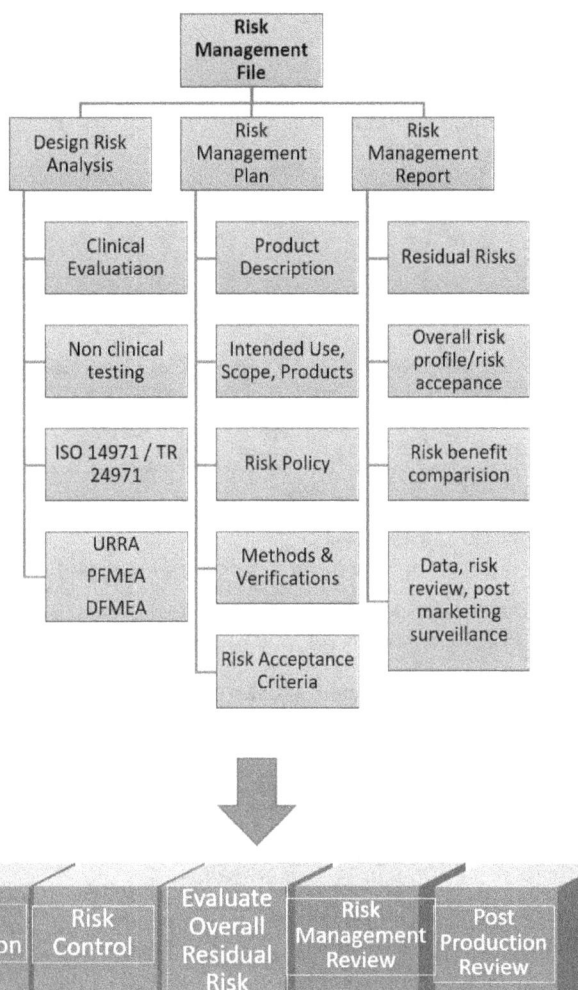

Overall Residual Risk

Benefit-risk analysis

The purpose of a Benefit-risk analysis is to document that residual risks are outweighed by the benefits of the device. When conducting the benefit-risk analysis, the criteria for acceptable must be taken from the risk management plan. Risks that are not acceptable and where additional risk control is not possible, the benefit-risk analysis must conclude that the benefits outweigh the residual risks.

Criteria of benefit-risk analysis

The benefit-risk analysis should take into regulatory factors but also the clinical and medical outcomes as a result of the availability of a device or product.

Hence, prior to commercial launch, clinical investigations may be required to determine that the balance between benefit and residual risk is acceptable and that the product is safe and effective with acceptable probabilities of occurrence of harm. Benefit risk analysis can be aided by comparative review of similar devices that are already on the market and review of generally acknowledged state of the art principles.

Residual Risk

After risk control and risk mitigation measures are applied, a new (follow up) risk assessment should completed to determine residual risks. If the residual risk is deemed unacceptable then further risk control measures should be applied.

Post Production Review

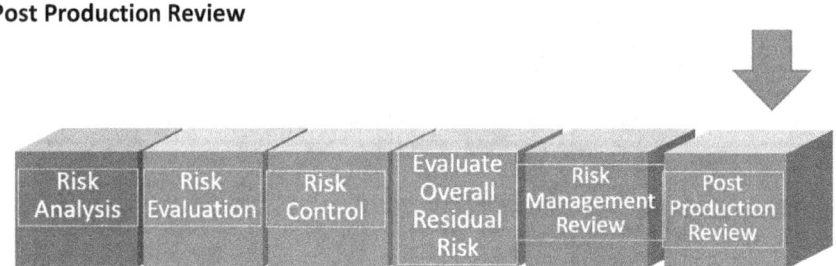

The goal of post-production risk management activities is to drive updates to the risk management file based on :

-Unanticipated risks or new or emerging risks identified

-Risks associated with product-use (use errors)
-Product performance levels relative to residual risk levels

-the likelihood of occurrence for hazards/harms change from previous established levels

Based on this information, the Risk Management report should be kept under review and updated if the risk profile is changed.

FMEA, Failure Mode and Effects Analysis

Failure Modes & Effects Analysis (FMEA) is a risk management analysis tool used during the design or to assess the manufacture of products to that can be used to manage risks caused by failure modes.

Types of FMEA

UFMEA assess the failure modes that occur during product-use and examines the robustness of product design, the intended use and also any reasonably foreseeable misuse by the end-user.

DFMEA assess the failure modes related to design of a system, product, feature, component of a final product and degradation of the product over its expected life.

PFMEA assess the failure modes that are related to the manufacture process of the product, including the safety of process operators.

Risk Management and Role of Standards

International standards such as ISO, ASTM, IEC are commonly used alongside ISO 14971 in regards to medical devices. These standards provide designers and manufacturers with information that has been reviewed by working groups and experts relevant to their industry. There publication is gated by peer evaluation, drafts that are subject to review and feedback and eventually subject to voting.

When performing risk management, the manufacturer first considers the medical device being designed, its intended use, its characteristics related to safety, and the associated hazards and hazardous situations.

Designers and Manufacturers can identify the product standards and process standards that contain specific requirements and help reduce the risks of use.

There are two distinct types of standards. (1) Product standards and Process standards. Both types play a part in delivering safe and effective medical devices, however, the product standards are more specific to the device in its scope.

Product standards can utilize the following:

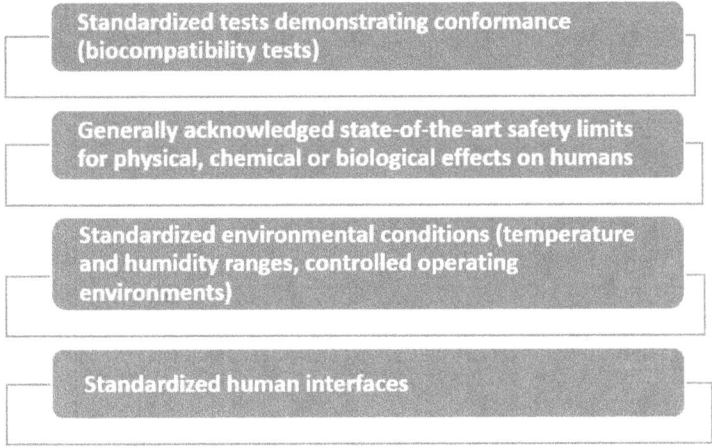

Process standards can be used in the following ways:

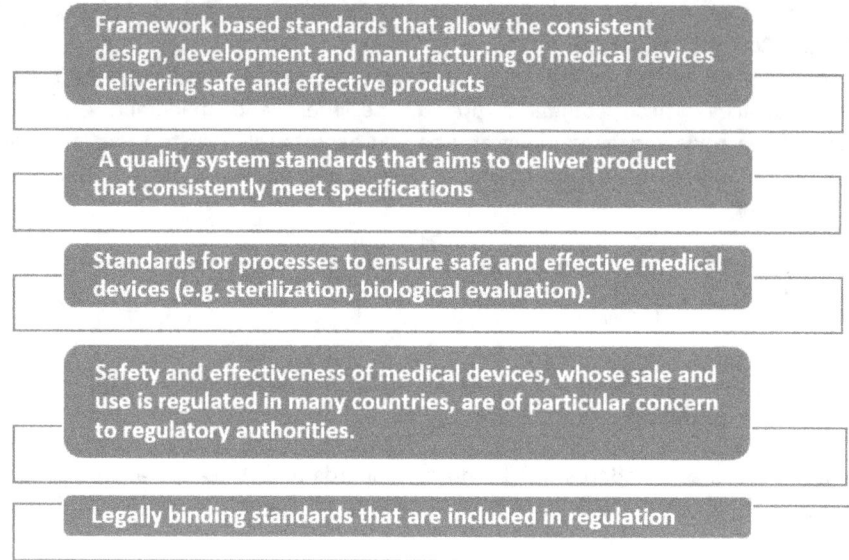

The following are examples of process standards applicable to manufacturers that can be used to assist in the Risk Management activities within a company:

ISO 16142-1 Essential Principles relating to Risk

This is a product safety standard and addresses safety and performance requirements that should be considered. Specifying the correct safety and performance requirements can result in successful risk reduction and risk acceptability.

ISO/IEC Guide 63, Guide to the development and inclusion of aspects of safety in International Standards for medical devices

This guide provides recommendations on the development and inclusion of safety requirements for international standards for medical devices. Their uses can be summarized as follows:

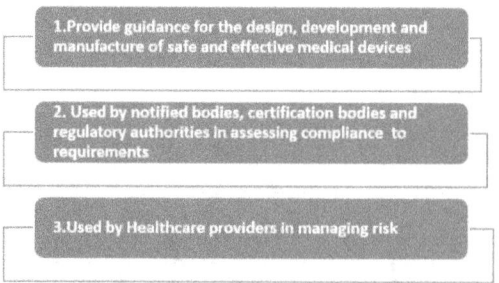

IEC 62366-1, Medical devices, Application of usability engineering

to medical devices

Risk Management has a number of inputs that help inform and assist in the application of the principles of risk management, e.g. identification of harms, hazardous situations- risk estimation/evaluations and risk analysis.

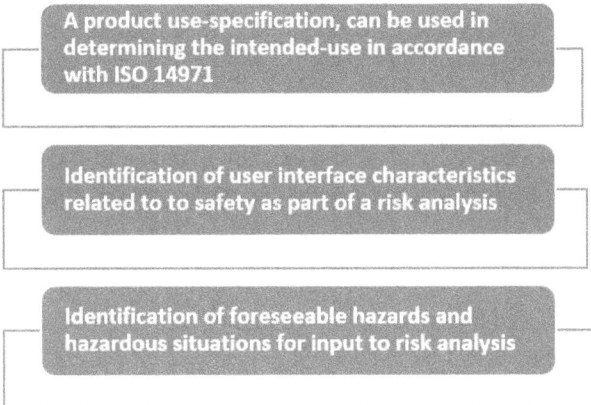

ISO 10993-1, Biological evaluation of medical devices - Part 1: Evaluation and testing within a risk management process

ISO 10993-1 covers the biological evaluation of medical devices within a risk management process, and takes into account the overall evaluation and development of each medical device.

In applies the same approach of ISO 14971 and involves:

(1) the identification of biological hazards for the medical device

(2) estimation and evaluation of the risks,

(3) the control of risks and monitoring the effectiveness of the risk controls

The biological evaluation utilizes the following information:

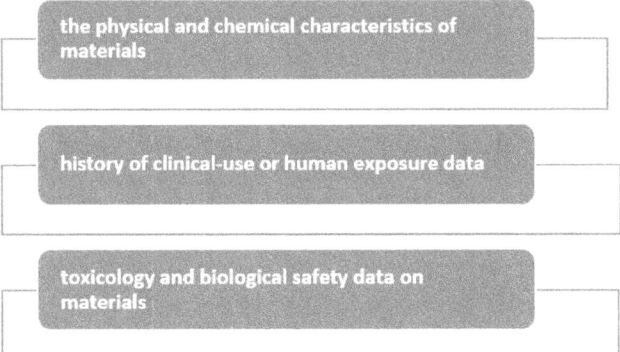

It is important to state that the Biological Safety Evaluation Report, is an element of the risk management file and may be referenced in the risk management report or other risk documentation, as required.

ISO 14155 - Clinical investigation of medical devices for human subjects — Good clinical practice

ISO 14155 is a standard that covers good clinical practice for medical devices for human subjects. It can be of use in determining the clinical risks and the benefit-risk analysis.

Usability Engineering and Medical Devices

Usability Engineering examines the format and function of user interfaces on medical devices and how they work to allow effective application by the user while studying the ease of user learning for a medical device.

Identification of Use Errors

The process of conducting Usability (Engineering) studies plays a key role in identifying scenarios where reasonably foreseeable misuse occurs.

Use Error is defined as a *"user action or lack of user action while using the medical device that leads to a different result than that intended by the manufacturer or expected by the user"*

Use Difficulty

Use Difficulties include repeated attempts to complete a task,
- ✓ hesitating,
- ✓ excessive "exploring" of the interface
- ✓ unexpectedly referring to the labeling information

Close Call

When a user makes a Use Error but then takes an action to "recover" and prevent the harm from occurring.

Success

Usability testing or usability engineering studies can be performed during the development of a new product. It acts as a verification that a device is designed appropriately and can identify scenarios or conditions that users could present a use error or usability risk to the patient or user.

IEC 62366-1 Medical Devices-Application of usability engineering to medical devices is referenced both in ISO 14971 and ISO/TR 24971. While Risk Management and Usability Engineering are separate processes, they both supplement and overlap in their intent.

As defined above, Use Error is defined as a *"user action or lack of user action while using the medical device that leads to a different result than that intended by the manufacturer or expected by the user"* Technical report, ISO 24971:2020.

This covers the following errors:
> -the inability of the user to complete a task.
> -Use errors resulting from a mismatch between the characteristics of the user, user interface, task, or use environment. Users may be aware or unaware that a use error has occurred.

Exception (to a use error):
- An unexpected physiological response of the patient is not by itself considered use error.
-A malfunction of a medical device

Identification of hazards from use errors
The usability testing or studies can highlight if issues occur when the device is used by the patient- for example, do people use the medical device in a way that it is not intended to be used or not in accordance with the instructions for use.

Hazards from reasonably foreseeable misuse
Some hazards and hazardous situations may be a result of reasonably foreseeable misuse. Engineering usability studies can also help identify and confirm reasonably foreseeable misuse scenarios.

Product Realization Process and Risk Management
Manufacturing companies that are also responsible production but design and development of medical devices are required to have processes and procedures in regards not only to risk management but also Product realization. Risk management and Process realization are normally separate processes with different procedures and SOPs. However, regulations e.g. EU MDR, require that the two processes work together with design and development taking into account risk management. Above all, this is to ensure safety requirements are included in development process and that risk are identified and tracked during the development lifecycle in order to ensure they are addressed or mitigated. The review of the results of the design verification activities during development to verify the risk controls were effective is part of this process.

FAILURE MODES AND EFFECTS ANALYSIS (FMEA AND FMECA)

Introduction

The Acronym FMEA otherwise known as Failure modes and effects analysis (FMEA) is a methodology of evaluating a product or process to identify the potential ways in which it may potentially fail. During the process of FMEA once the potential failures are identified- then the effects of the mode of failure on safety, efficacy, performance of of the product or indeed the process (or both).

FMEA can be rolled out to hardware, software, processes including human action, and their interfaces, in any combination.

Business and Quality Case for FMEA

Why in simple terms FMEA is a risk management risk assessment tool it also can be used to improve business metrics, reduce waste, reduced we rework, improve quality, and from a safety perspective reduce or prevent harm, injury or adverse events from occurring.

The process of completing a female start with identifying the process steps. Each process step can be reviewed to identify potential failures that can lead to hazards are hazardous situations. Once these are identified the next step is produce the likelihood of failures, eliminate them if possible by design, and also reduce the effects are the consequences of the failure mode. The cause of the failures can also be documented. Taking a business viewpoint can bring an awareness of the effect off failures on yield, downtime, scrap, cost and customer satisfaction. For quality, and the safety and performance of the finished product, FMEAs for medical devices need to cover the potential harm to patients as a result of hazards and hazardous situations.

Why FMEA?

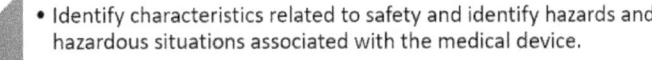

1 • Identify characteristics related to safety and identify hazards and hazardous situations associated with the medical device.

2 • Identify hazards and hazardous situations that are completely covered by the international product safety standard.

3 • ensure the design specifications of the medical device conform with the requirements in the standard that serve as risk control measures.

4 • Verification of the implementation of the risk control measures for hazardous situations is obtained from design documentation review

5 • Verification of the effectiveness of the risk control is demonstrated and meets internatonal product safety standards

Methodology for FMEA

There are 3 broad phases of conducting a FMEA which are:

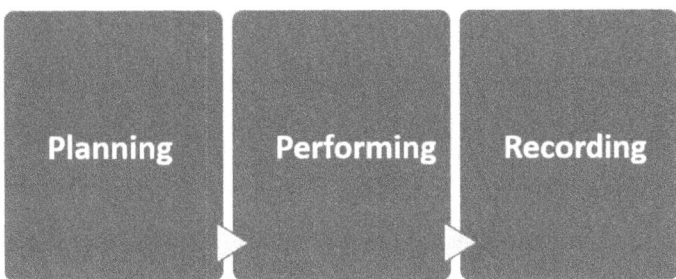

The creation of a FMEA is often an iterative process- meaning it can go through a number of drafts and revisions as it is populated and more information becomes available and risk analysis and estimation is completed

The FMEA will, however, always identify the effects of failure modes on the top level of the hierarchy within the analysis scope.

In the planning stage, the boundaries of the FMEA should be established. This can be helped by creating a process flow diagram. Some process steps or sub process may be outside the boundary or scope of the FMEA to somewhat make the scope and size of the FMEA manageable or if a separate and independent FMEA is more suited.

Planning an FMEA involves considering why an analysis is to be performed, what equipment or process elements. What is the goal of the FMEA, is it an overarching FMEA or should it focus on each step of a process. After consideration and planning of a FMEA the following information should be understood.

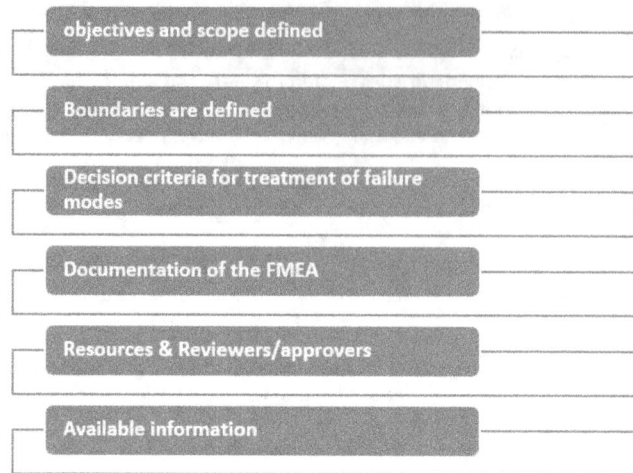

Definition of decision criteria for treatment of failure modes

Decision criteria for the treatment of failure modes helps to focus or prioritise on F.Ms that are the most critical. For medical devices, safety related failure modes, or failure modes that may impact the patient must be treated as critical. A more detailed priority may be established based on the severity of the failure mode. Some failure modes will result in risk that is unacceptable.

Criticality of failure modes can also be assessed by:

• reviewing the severity of the failure effect (on patient, product or process)

• the likelihood that the failure mode might occur and lead the consequence or harm

• detectability of the failure mode

Based on Severity, Occurrence, Detection scores (SXOXD) an RPN, Risk Priority number can be established.

Identify Failure Modes

Failure modes identification:

DFMEA

- o If similar devices exist, failure modes may be known or developed based on existing data and knowledge.
- o Failure modes may present themselves in different modes of operation
- o Failure modes may present themselves at different stages of the product lifecycle
- o Can failure modes be a result of storage or transport
- o Is there material issues that can result in failure modes

PFMEA
- o If a similar or existing manufacturing process is used, the process failure modes may be available
- o Is the process subject to environmental changes or trends

o   Are there failure modes due to operational issues, human error or automation issues

Identify existing controls or detection methods

After failure modes are identified for each process step or element, the existing control measures and any detection methods should be recorded against each failure mode. Controls may prevent a failure mode or reduce its occurrence while detection allows identification of the failure mode which allows reaction or intervention.

Identify effects of failure modes

A failure effect can be understood as a consequence of a failure mode. Failure effects may be caused by one or more failure modes. The description of each failure effect have a level of detail that allows the assessment of the severity level and what the consequences would be.

Identify failure causes

The cause of the failure and how it occurs is helpful in reducing the likelihood of failure or its consequences.

Common cause and common mode failures

Common cause failures happen when more than one element fails at the same time or within a short period of time that result in the effect of the failures.

Elements can also fail in the same way or with the same failure mode- however, this can be due to different causes or the same cause.

The likelihood of occurrence can be estimated using:

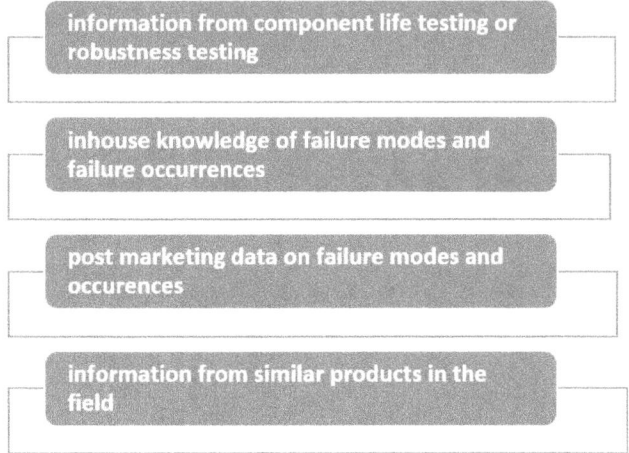

Identify actions

This could include additional controls, revalidated, additional audit requirements of engineering changes.

If new controls or detection methods are agreed as necessary and introduced, re-analysis should be completed post implementation of changes. This is to assess if:

• any new failure modes or effects have been introduced (e.g. re-sterilisation may damage packaging ; and

• the criticality of the particular failure modes is deemed acceptable.

7.0 DATA INTEGRITY AND PRINCIPLES OF COMPLIANCE

Introduction

Data generated by or used in GxP impacting activities must be handled and protected in accordance with international and national regulatory requirements. The application of data integrity applies to many industries and products that touch the lives of patients and end users across the globe. Some examples of products that must meet data integrity regulations include (1) active pharmaceutical ingredients, (2) medical devices, (3) medicinal products, (4) vaccines and (5) cosmetics.

The below agencies and regulatory authorities provide specific requirements on data integrity:

➢ EU GMP – EudraLex – Rules Governing Medicinal Products in the European Union Volume 4 – Guidelines to Good Manufacturing Practice for Medicinal Products for Human Use – Products for Human and Veterinary Use, Annex 11: Computerised Systems – (1, 7.2, 17)

➢ FDA – 21 CFR Part 11 – Food and Drug Administration – Electronic Records; Electronic Signatures – Scope and Application (C)

➢ FDA- 21 CFR Part 211 – Food and Drug Administration – Code of Federal Regulations - Good Manufacturing Practices - 211.188a, 211.194.2, 211.194.8

➢ ICH E6 – International Conference on Harmonisation - Guideline for Good Clinical Practice
(5.2.1, 8.1, 8.3)

➢ MHRA – United Kingdom - Medicines and Healthcare Products Regulatory Agency - GMP Data Integrity Definitions and Guidance for Industry (2015)

➢ PIC/S Guidance PI 011-] – Pharmaceutical Inspection Convention Scheme - Good Practices for Computerised Systems in Regulated "GXP" Environments

Within the life science industry the saying goes "if it's not written down, it didn't happen". This is a powerful message that is a suitable starting point for data integrity. In the current and present day, the mere mention of data integrity quickly conjures an image of Excel sheets, big data, databases and computers in our minds. However, it has a broader impact with its roots in the basics of good science – good documentation.

Data integrity indeed does apply to "soft" or electronic data but also applies to paper-based systems and records. GxP is the umbrella acronym that stands of "good practices" in all our tasks and activities, be it laboratory testing, process engineering and so on. A core element in meeting GxP is abiding by "Good Documentation Practices" (GDP). Having good written records is fundamental to patient and product safety within the pharmaceutical, biopharmaceutical and medical devices industries. So, data integrity begins with the small stuff — real-time data collection, real-time review, honest and accurate recording of data and events.

The integrity of data relies on several factors. It can be influenced by a company's culture or approach to doing business. It can also be affected by the level of experience or knowledge within a company. Many traditional engineering companies outside the regulated life science community simply do not have the need to be so thorough in their handling of data and information. Within a GxP environment, controls, training and the design and operation of systems and processes influence data integrity on a day-to-day basis. Most of the time, those affected by the controls or systems do not think of them, but they can either support or inhibit data integrity and the reliability of data. Obviously, equipment, systems and processes should play a key role in making data reliable and accurate.

Configuration Identification

Software and hardware packages should be identified by a unique product identifier and a version number. For the software end-user, the parts of an automated system that are subject to configuration management should be clearly identified. The system should therefore be broken down into configuration items. These should be identified at an early phase of development so that a complete list of configuration items is defined and maintained. The application-specific items should have a unique name or version ID. The depth of detail when specifying the elements is decided by the needs of the system, and the organisation developing that system.

Requirements for the User ID and Password

User ID: The user ID of a system should have a minimum length agreed with the customer and should be unique within the system.

Password: A password should always consist of a combination of numeric and alphanumeric characters. When setting up passwords, the number of characters and a period after which a password expires should be stipulated. The structure of the password is normally selected to suit the specific customer. The configuration is described in the security settings section of password policy. Criteria for the structure of a password are as follows:

> Minimum length of the password
> Use of numeric and alphanumeric characters

➤ Case sensitivity

Audit Trail

The audit trail is a control mechanism of a system that allows all data entered or modified to be traced back to the original data. A reliable and secure audit trail is particularly important in conjunction with the creation, change or deletion of GMP-relevant electronic records. In this case, the audit trail must archive and document all the changes or actions made along with the date and time. Typical contents of an audit trail must be recorded and describe the procedures "who changed what and when" (old value/new value).

Data: any data (numerical or otherwise) which is collected or processed as part of GxP activities in order to generate GxP documents and records using a paper-based or electronic process.

Data Handling: Any GxP task that involves creation, entry, review, approval, analysis, reporting, storage, archival, retrieval, or disposal of GxP data.

Data Integrity: The degree to which a collection of GxP data is managed through effective organisational, operational, and technical mechanisms to ensure GxP data reliability.

Data Life Cycle: Starts from the time of data creation to the point of use and during its retention, archival, retrieval and eventual disposal

GxP Impacting: Any action that can impact the quality or safety of a product or critical process.

Application: Software installed on a defined platform/hardware providing specific functionality.

Bespoke/Customised Computerised System: A computerised system individually designed to suit a specific business process.
Commercial Off-the-Shelf Software: Software commercially available, whose fitness for use is demonstrated by a broad spectrum of users.

IT Infrastructure: The hardware and software such as networking software and operation systems, which makes it possible for the application to function.

Life Cycle: All phases in the life of the system from initial requirements until retirement including design, specification, programming, testing, installation, operation and maintenance.

Process Owner: The person responsible for the business process.

System Owner: The person responsible for the availability, and maintenance of a computerised system and for the security of the data residing on that system.

Third Party: Parties not directly managed by the holder of the manufacturing and/or import authorisation.

The Life Cycle of Data

Regulations that speak to GxP and data integrity can apply to many different streams within the life science sector as previously mentioned. From medical devices to pharmaceuticals, all act in different manners, with long and short term applications. Take the example of a total knee replacement. Many designs now ensure their effectiveness in excess of ten years, even up to twenty years depending on individual circumstances. This requires many key records within manufacturing to be kept for several decades. Thus, data retention requirements specify the retention periods of such documents. The integrity of GxP data must be protected during the entire data life cycle, from creation of the data and records to the eventual destruction of data after the retention period is fulfilled. Data integrity equally applies to:

> ➢ Equipment
> ➢ Computerised systems
> ➢ Test records
> ➢ Inspection records
> ➢ Material certificates

Data integrity ensures that patient safety, product quality, and product supplies are generated by the product life cycle processes.

Process Design

Failure to maintain data integrity can occur throughout the life cycle of data; however, a thoughtful design of systems can prevent breaches in data and restrict the severity of any attempts to alter data. Therefore, design should aim to include controls and preventative measures. At a high level, this can be achieved by:

> ➢ Limiting access to GxP events and data
> ➢ Standard Operating Procedures (SOPs)
> ➢ Training
> ➢ System owners

Data Reliability

Data reliability is the foundation to achieving cGxP data integrity. The FDA's ALOCA model can be used to enforce data reliability.

Accuracy: the GxP data is recorded, calculated, analysed, and reported as found and correctly.

Attributable: any actions or calculations performed on GxP data can be attributed to or traceable to the person that performed the actions and the date and time at which they were performed.

Legible: the GxP data is recorded in a clear and human-readable form.

Contemporaneity: the GxP data is recorded at the same time as the observation/measurement is made or as soon as possible after the event.

Original: the initial data recorded is available and not altered.

An additional point to make it that of trustworthiness. It is assumed that engineers and scientists etc. working across the life science industries are ethical and do not falsify data or information. Typically companies can implement a code of practice or ethical behaviour programme to desist people from intentional unethical behaviour or the falsification of records.

Data Creation: The point at which the values or data is created. The data and information is original (raw).

Data Authentication: Within a GxP environment, authentication refers to the approval of data (electronic signatures). E-signatures are key controls within software that prompt the user to enter a unique username and password to acknowledge a recording or action. The e-signature should create a permanent link with the electronic record that cannot be removed and can be viewed through an audit trial.

Data Protection: Once the data is created, the handling of the data must ensure data integrity. For electronic data, this includes access control to computer systems. Other practical restrictions can also be made such as limiting room and site access to authorised personnel.

Data Retention: This refers to the controlled storage, backup and arching of data. Retention of records may be required for several decades depending on the type of data and the regulatory requirements relating to the particular product or industry.

Technical Controls

The benefits of modern software and computerised systems allow robust and complex data handling and calculations to be completed. With this modern capability that is becoming more powerful comes more responsibility with regard to the use of data.

The computerised systems used to generate, gather or interpret GxP data must fulfil several criteria. First and foremost, they must be fit for the intended use. The software and hardware must be validated and proven to be consistent and reliable. Some general considerations for the use of computerised systems include:

> Systems designed to foster integrity of GxP data
> User requirements specification detailing the intended use and required functionality
> An approved vendor with certification to ISO 9001 or other quality management standards
> Software should meet the requirements of regulations such as FDA 21 CFR Part 11
> Written procedures on how automated processes function

It should not be an easy process for personnel to alter or corrupt data when using computerised systems. GxP-impacting computer systems should have controls that prevent unauthorised access along with audit trail history.

Audit trail design and configuration capture key critical processes, events, settings and information. This enables any investigations of quality events impacting data integrity to be reviewed and analysed.

☐

Computer System Design and Development

For computer systems, software requirements are typically stated in functional terms and are defined, refined and updated during the development phase. Success in accurately and completely documenting software requirements is a crucial factor in successful validation of the resulting software. A specification* is defined as "a document that states requirements." It may refer to or include engineering drawings or other relevant documents *21 CFR 820.3(y).

There are different kinds of written specifications:

> - User requirements specifications
> - System requirements specification
> - Software requirements specification
> - Software design specification
> - Software test specification
> - Functional design specification

All of these documents establish "specified requirements" and are design outputs for which various forms of verification or validation are required. The URS must also define non-software requirements and hardware. Non-functional requirements such as maintainability and usability can also be included. There should be a clear distinction between mandatory regulatory requirements and optional features. Proper definition at this stage ensures the system meets data integrity requirements and prevents costly updates down the line.

Practical Elements to Data Integrity

Facilities and systems must be configured in a way that encourages compliance with principles of data integrity. Examples include:

> - Availability of clocks for recording times.
> - Access points to allow swift reference to GxP records at locations where tasks are completed.
> - Control of raw data.
> - Control of approved documents.

Organisational Controls

Regulated companies such as medical device, pharmaceutical and biotechnology companies are required to operate under a quality management system. For medical devices, ISO 13485 serves as a quality management system. Likewise, the FDA Code of Federal Regulations 21 CFR Part 211 functions as a QMS for finished pharmaceuticals.

Organisational controls for Data Integrity can address:

➤ Assessment of GxP computerised systems
➤ Management of GxP computerised systems
➤ Electronic Records Implementation and handling
➤ Use of Electronic signatures
➤ Quality Risk Management

Operational Factors

Operational factors refer to process or manufacturing errors, deviations or non-compliance to established procedures that may impact data integrity.

GxP data handling activities should be designed to limit human intervention. As with human intervention there can be errors or omissions. Furthermore, it may call into question the reliability of the data. Mistake-proofing methodologies should be developed to avoid human error related breaches in data integrity. As with any system or technology, training is a fundamental step. Building upon training, exposure to GxP data systems and on-the-job training all play a part in delivering a system that is robust and meets regulatory requirements. It is important to remind ourselves that while regulations are the driving force to comply with data integrity, the ultimate goal is always the protection and safety of the patient or end user of the product, medicine or treatment.

Software Validation

Where there is the potential to affect product conformance to requirements or where software or IT systems provide support to aspects of quality management, validation is required. Most companies categorise software validations to account for the different applications of software and IT systems. For example, enterprise systems, such as the drawing package SolidWorks would be validated in a different manner to manufacturing systems that contain software (a.k.a. embedded software).
"Embedded" software is where the software is integrated into the manufacturing equipment. Embedded software is typically validated during the equipment qualification stage, process validation stage or test method validation. Enterprise software falls outside of equipment or process validation but does require validation if it impacts product quality or is used to make quality decisions. Standalone systems such as ERP (Enterprise Resource Planning) systems also require validation.

Software Validation and GAMP

Good Automated Manufacturing Practice (GAMP) is a set of guidelines for manufacturers and users of automated systems in regulated industries. GAMP specifically impacts the medical device, pharmaceutical and biopharmaceutical industries. The application of GAMP and validation of automated systems in manufacturing helps ensure that regulated medical devices and medicinal products have the required quality and are manufactured according to good practices, meet regulatory and legal requirements and ensure patient safety. GAMP ensures quality is in-built into each stage of the manufacturing process. Therefore, GAMP has a place in all aspects of automation and production, including the handling of raw materials, control of facilities and equipment etc.

Automated System: Term used to cover a broad range of systems, including automated manufacturing equipment, control systems, automated laboratory systems, manufacturing execution systems and computers running laboratory or manufacturing database systems. The automated system consists of the hardware, software and network components, together with the controlled functions and associated documentation. Automated systems are sometimes referred to as computerised systems; in this guide the two terms are synonymous.

Commercial Off-the-Shelf (COTS): Configurable programs and stock programs that can be adapted to specific user applications by "filling in the blanks" without (COTS) altering the basic program.

Computer System Validation: A process that confirms by examination and provision of objective evidence that the computer system conforms to user needs and intended uses. System validation is a process for achieving and maintaining compliance with GxP regulations and fitness for intended use by adoption of life cycle activities, deliverables, and controls.

GAMP 5: A set of guidelines that offers a risk-based approach to ensuring the compliance of GxP-impacting computerised systems.

V- Model: A development process which sets out a roadmap of stages and deliverables during a project.
21 CFR Part 820: FDA requirements pertaining to medical devices.

User Requirement Specification, URS: The URS is a critical document that defines the requirements of the computerised system and agreement to the requirements.

Software Requirement Specification, SRS: An SRS can be written to interpret the requirements of a URS and how they relate to the requirement or how the requirement is met in practical terms regarding software.

Functional Design Specification, FDS: A functional design specification is a document that specifies how particular requirements are met — this can be a combination of how the equipment/process operates mechanically/automatically etc. An FDS is typically written in response to a URS.

Computer System Validation Life Cycle

The computer system validation life cycle refers to all activities from initial concept to retirement of a computer system. The life cycle of the system includes the defining of, and performance of activities in a systematic way from conception, requirements, development or configuration, testing, release and operational use. The four GAMP life cycle phases include:

- ➤ Concept
- ➤ Planning and project stage
- ➤ Operation
- ➤ Retirement

The concept stage is concerned with understanding the need or the problem to be addressed. We will see that the user requirement specification (along with other specifications) and the initial risk assessment help to drive a project forward in a systematic manner. The most common life cycle approach for computerised and automated systems is the V-Model. The GAMP-based V-model lays out a roadmap which facilitates the validation of equipment and automated systems.

The planning and project stage involves the planning of the validation effort required to implement the system into the business area(s) based on identification and approval of system concept. This phase includes assessments of the regulatory and system risks, supplier assessment, development of validation strategies, identification of deliverables that will be generated, definition of the business process the system will support as well as the user requirements which the system will fulfil.

Design, development   and configuration of the hardware and software is also required to meet the system requirements as per specifications. In the case of custom software components, this effort could also include detailed software design and developmental testing to ensure readiness for verification testing.

The verification stage confirms that specifications have been met and releases the system for use. This phase will involve multiple stages of reviews and testing depending on the system type, the development method applied and its use. Once verification activities have begun, any changes to the system must be captured through change control.

On successful completion of the verification activities, the system is then released for effective use. The test strategy and other verification activities will vary widely between simple equipment and more complex customised/configurable systems. The verification and validation approach is typically agreed and detailed at the validation planning stage. The VP can be updated accordingly as the project develops with more detail being added. Alternatively, a test strategy document or matrix could be written to provide more specific test plans.

Verification deliverables vary based on the complexity and level of customisation of the system in question. Corporate or company specific procedures also shape the required activities to be completed and reported. Some generic deliverables are listed below.

- ➤ Approval, execution and review of test protocols
- ➤ Writing and approving SOPs for operation and maintenance of the system
- ➤ Traceability matrix

- ➤ Completion of any risk mitigations (e.g. updates to FMEA etc.)
- ➤ Validation summary report(s)

Validation reporting requirements vary depending upon the scope of the system and should also be driven by a procedure and template. The validation plan can also outline the deliverables and what needs to be addressed in the report. A Validation Summary Report (VSR) should be written to summarise the results of executing the VP, the documents created for the validation activities and the testing performed. Finally, the VSR indicates the acceptance of the system/equipment by the user and the project team and states that the equipment is released for commercial operation/production.

The operation phase supports the need to maintain compliance and fitness for intended use after the system is released for normal use. It is important to ensure the system remains within a continued validated state. All proposed or necessary changes to the system must be assessed and controlled as part of a change control process. Once the system has been accepted and released for use, the operation phase begins. This phase consists of maintaining the system's compliant state and fitness for intended use through the control of the procedures supporting the system's operational use.

During the operation phase, the below activities are typically completed:

- ➤ Ongoing training
- ➤ Preventative maintenance
- ➤ Service management and performance monitoring.
- ➤ Change control
- ➤ Periodic review
- ➤ Maintaining system security
- ➤ Records management
- ➤ Calibration

The retirement phase involves the planning and proper management of activities relating to the removal of systems from service (shutdown). The retirement should take into account the storage of any data and any data migration that needs to occur prior to retirement. The retirement plan, if needed, will outline the retirement strategy from the roles and activities that will be conducted to the removal of the system for use. A retirement summary report is produced that documents the results of the activities defined in the retirement plan including:

- ➤ Retirement plan and timelines.
- ➤ Summaries of any data migration activities.
- ➤ Identification of the storage location of documentation relating to the system.
- ➤ Obsoleting of SOPs.

It must be stressed that GAMP is a set of principles, a set of guidelines that aim to achieve compliant computerised systems that are fit for intended use. GAMP guidelines differ to 21 CFR QSR regulations as they are not legal or statutory requirements. However, they represent industry best practice and complement the validation efforts that are legal requirements and statutory requirements.

Regulatory Review

Software validation is a requirement of the quality system regulation, 21 Code of Federal Regulations (CFR) Part 820. Validation requirements apply to:

(1) software used as components in medical devices,
(2) software that is itself a medical device, and
(3) software used in production of the device or in implementation of the device manufacturer's quality system.

Note: EU GMP Annex 11, provides information on the inspection of 'Computerised Systems'.

In addition, computer systems used to create, modify, and maintain electronic records and to manage electronic signatures are also subject to the validation requirements. Such computer systems must be validated to ensure accuracy, reliability, consistent intended performance, and the ability to discern invalid or altered records. The regulated user should be able to demonstrate through the validation evidence that they have a high level of confidence in the integrity of both the processes executed within the controlling computer system and in those processes controlled by the computer system within the prescribed operating environment.

System Categorisation

GAMP 5 makes provision for four categories of software in order to distinguish the level of tcustomisation/configurability that exists across software serving different functions:

GAMP Software Category 1, Operating Systems
GAMP Software Category 2, Non-configured software
GAMP Software Category 4, Configurable software packages
GAMP Software Category 5, Custom Software

GAMP Software Category 1, Operating Systems

Category 1, operating systems, covers established commercially available operating systems. These systems are not subject to validation themselves. The name and version of the operating system must, however, be documented and verified during Installation Qualification (IQ). Application software hosted on operating systems needs to be validated.

GAMP Software Category 3, Non-Configured Software

Category 3 covers commercially available, standard software packages and "off the-shelf" solutions for certain processes. The configuration of the software packages should be limited to adaptation to the runtime environment (for example network and printer connections) and the configuration of the process parameters. The name and version of the standard software package should be documented and verified in an installation qualification (IQ). Special user requirements, such as security, alarms, messages, or algorithms must be documented and verified in an operational qualification (OQ).

GAMP Software Category 4, Configurable Software Packages

GAMP Software Category 4, Configurable Software Packages Category 4 covers configurable software packages that allow special business and manufacturing processes. This involves configuring predefined software modules. These software packages should only be considered as belonging to Category 4 if they are well-known and mature. Normally, a supplier audit is necessary. If this is not available, the software packages should be handled as Category 5. The name, version, and configuration should be documented and verified in an installation qualification (IQ). The functions of the software packages should be verified in terms of the user requirements in an operational qualification (OQ). The validation plan should take into account the life cycle model and an assessment of suppliers and software packages.

GAMP Software Category 5, Custom Software

GAMP Software Category 5, Custom Software Custom/Bespoke Software (GAMP Software Cat 5) is software that contains custom code designed or modified specifically for a particular customer. As the code is custom, it presents a greater risk. This risk must be mitigated with the right approach to the validation.

GAMP Considerations

Correctly assigning a GAMP software category to equipment, systems or processes is an important activity that should be completed early on in the planning stage of a project. There must of some degree of familiarity with the equipment or system. The manufacturer or vendor can be a source of information that may help the designation. In many cases, companies create tools or processes that help determine what GAMP software category applies. These have different names such as questionnaires, screening tools, planning tools etc.

Planning Stage

Initial Impact/Risk Assessment – takes place during the planning phase to identify the level of impact and GxP relevance of the system/equipment. (Tools used: High Level Risk Assessment).

Specification Stage

Functional or Quality Risk Assessment – takes place during the specification phase and identifies potential risks and possible mitigations to be to be introduced to the process. (Tools used: Quality Risk Matrix, (p)FMEA).

Changes to the System

Impact Assessment of Changes – takes place as part of the change control process in the system operational phase.

Quality Risk Matrix

A QRM is a risk assessment that identifies and manages the risk to patient safety, product quality and data integrity that relate to system processes. Risk scenarios or potential causes should be developed for each identified function or process step and then assessed for the impact on patient safety, product quality or data integrity. Risk mitigations and controls should then be introduced to address both medium and high levels of risk. The QRM requires three "assessments" in order to produce an estimation or overall risk (low, medium, high),

- ➢ Assess likelihood
- ➢ Assess detectability
- ➢ Assess severity

Traceability Matrix

A traceability matrix should be prepared as required in accordance with company and internal policy. It is also recommended by GAMP guidelines, ASTM E2500 and ISPE risk-based approaches to validation. The matrix links the user requirements and specifications to testing and validation activities. A traceability matrix illustrates that all user requirements are traceable to the verification/validation activity or vendor documents as relevant (FDS if applicable, design specifications etc.) Generally, individual organisations will have an approved template to work from. However, the URS structure can form the basis of the template, with additional columns added to document the test/verification method or reference documents (such as FDS and vendor specifications and design documents)

☐

21 CFR Part 11

This section specifically covers the regulatory requirements of part 11 of Title 21 of the Code of Federal Regulations; Electronic Records; Electronic Signatures (21 CFR Part 11). Part 11 of the FDA CFR is relevant to "records in electronic form that are created, modified, maintained, archived, retrieved, or transmitted under any records requirements set forth in agency regulations."

As of 2007, several sections of the regulation have been identified as excessive and the FDA announced in guidance that it will exercise enforcement discretion on some parts of 21 CFR part 11. This has been welcomed by some manufacturers but it has also caused a degree of confusion. The requirements relating to access controls are the most fundamental requirements and are routinely enforced. The "predicate rules" that required organisations to keep records in the first place are still in effect. If electronic records are illegible, inaccessible, or corrupted, manufacturers are still subject to those requirements.

If a regulated firm keeps "hard copies" of all required records, those paper documents can be considered the authoritative document for regulatory purposes. This then means that the computer system is not in scope for electronic records requirements, although subject to predicate rules which still require validation. If the "hard copy" is to be identified as the authoritative document, the "hard copy" must be a complete and accurate copy of the electronic source. The manufacturer must use the hard copy (rather than electronic versions stored in the system) of the records for regulated activities.

Definition of Records

The FDA has deemed the following records or signatures in electronic format subject to 21 CFR part 11:

Records that are required to be maintained under predicate rule requirements and that are maintained in electronic format in place of paper format. On the other hand, records (and any associated signatures) that are not required to be retained under predicate rules, but that are nonetheless maintained in electronic format, are not part 11 records. Records that are required to be maintained under predicate rules, that are maintained in electronic format in addition to paper format, and that are relied on to perform regulated activities.

Records submitted to FDA, under predicate rules (even if such records are not specifically identified in agency regulations) in electronic format (assuming the records have been identified in docket number 92S-0251 as the types of submissions the agency accepts in electronic format). However, a record that is not itself submitted, but is used containing nonbinding recommendations in generating a submission, is not a part 11 record unless it is otherwise required to be 205 maintained under a predicate rule and it is maintained in electronic format.

Electronic signatures that are intended to be the equivalent of handwritten signatures, initials, and other general signings required by predicate rules. Part 11 signatures include electronic signatures that are used, for example, to document the fact that certain events or actions occurred in accordance with the predicate rule (e.g. approved, reviewed, and verified).

The above definitions are taken from the FDA guidance document entitled "FDA Guidance for Industry: 21 CFR Part 11 - Electronic Records and Electronic Signatures: Scope and Application, August 2003." This document also provides recommendations on documenting key decisions that may be taken in relation to 21 CFR Part 11 applicability and compliance.

Requirements and Specifications

The need for compliance to 21 CFR depends on the type of technology and level of automation and computerisation involved in the manufacturing process or other actives that are GxP-impacting. Does the system store electronic records? Does the system require a login? Is there an audit trial? If a complex system is to be procured, the requirements need to be communicated to the manufacturer as part of a user requirement specification and/or software requirement specification.

General Guidance on Requirement Specifications

While the quality system regulation states that design input requirements must be documented, and that specified requirements must be verified, the regulation does not further clarify the distinction between the terms "requirement" and "specification." A requirement can be any need or expectation for a system or for its software. Requirements reflect the stated or implied needs of the customer, and may be market-based, contractual, or statutory, as well as an organisation's internal requirements.

There can be many different kinds of requirements (e.g., design, functional, implementation, interface, performance, or physical requirements). Software requirements are typically derived from the system requirements for those aspects of system functionality that have been allocated to software. Software requirements are typically stated in functional terms and are defined, refined, and updated as a development project progresses. Success in accurately and completely documenting software requirements is a crucial factor in successful validation of the resulting software. *Page 6 Guidance for Industry and FDA Staff General Principles of Software Validation A Specification* is defined as "a document that states requirements." (21 CFR 820.3(y)). It may refer to or include drawings, patterns, or other relevant documents and usually indicates the means and the criteria whereby conformity with the requirement can be checked.
There are many different kinds of written specifications, e.g., system requirements specification, software requirements specification, software design specification, software test specification, software integration specification, etc. All of these documents establish "specified requirements" and are design outputs for which various forms of verification are necessary.

Validation of Computerised Systems

The requirement for computerised systems to be compliant to 21 CFR part 11 needs to be identified early on in the project to ensure that the vendor or supplier of the systems or equipment can develop and build a system that meets the requirements of 21 CFR part 11. Computer system validation can be divided into three distinct phases: (1) planning, (2) design and development, (3) verification and (4) retirement.

Planning: This phase involves the planning of the validation effort required to implement the system and identification of key milestones and requirements. It requires supplier assessments, assessments of the regulatory and system risks, supplier development of a validation approach and the identification of deliverables that will be generated to support the implementation and operation of the system.

Design and Development: This phase consists of the design, development and configuration of the hardware and software required to meet the system requirements. In the case of custom software, design and developmental testing is important to ensure proper functionality prior to verification testing.

Verification: This phase confirms that requirements and specifications have been met. Testing is required to ensure the system operates as intended. Upon successful testing and verification, the system can be released for use. Once verification activities have begun, any changes to the system must be managed through change control. In case of successful completion of the verification activities (i.e. any deviation has been evaluated and addressed), the system is released for effective use. The operation phase supports the need to maintain compliance and fitness for intended use after the system is accepted and released for use.

Retirement: This phase consists of the planning, executing and summarising of the events required for system shutdown. It includes the appropriate handling of the supporting documents and the data contained within the system. While described here as a separate phase, a system's retirement can be handled as part of a new system implementation or as a separate project. Best practice when it comes to computer system validation is to adopt a life cycle approach which requires the completion of activities in a systematic way from system conception to retirement. Life cycle activities could be scaled according to system impact on product quality, patient safety and data integrity, system complexity and novelty, supplier assessment and business risk.

Definitions

Computer System: A computer/automated system consisting of the hardware, software, and network components, together with the controlled functions (personnel, procedures, and equipment) and associated documentation.

Computer System Validation: A process that confirms by examination and provision of objective evidence that the computer system conforms to user needs and intended uses. Computer system validation is a process for achieving and maintaining compliance with GxP regulations and fitness for intended use by adoption of life cycle activities, deliverables, and controls.

GxP-Regulated Computer Systems: Computer systems determined to have a potential impact on product quality, patient safety and data integrity; these systems are required to comply with the relevant GxP regulations.

Data Integrity: The degree to which data is reliable and without error. Data must be accurate, attributable, contemporaneous, original, legible and available. A breach of data integrity occurs when any person manipulates or distorts data and submits the results of that data as valid.

Predicate Rules: A predicate rule is any FDA regulation that requires companies to maintain certain records and submit information to the agency as part of compliance.

To gain a better understanding of the validation of computerised systems, consult the following publication: "FDA's Guidance for Industry and FDA Staff General Principles of Software Validation." See also industry guidance such as the GAMP 5 guide issued by ISPE for a useful reference.

Electronic Records

When it comes to the regulated industries such as the medical device industry, every process and procedure must be documented. Documentation ensures that everyone is working in the same manner with the same procedures. However, documentation is more than just writing down procedures and processes. It is also concerned with how documents are controlled, how they are updated and how they are stored.

Electronic Document management systems

Electronic document management systems aka EDMS are now the norm and gold standard for most medium to large organisations. Many companies that provide medical device manufacturers with an EDMS that can be customised to match the business processes particular to an organisation. With configurable or customisable software, validation and proper verification is important to ensure the system operates as intended. There are also regulatory requirements that stipulate the expectations and requirements of such systems. For example, the application of electronic signatures and the presence of audit trials. FDA 21 CFR Part 11 details the requirements with regard to electronic records and electronic signatures. For medicinal products in Europe, GMP V4 Annex 11 specifies similar requirements.

Record Retention

With regard to the part 11 requirements for the protection of records to enable their accurate and ready retrieval throughout the records retention period (11.10 (c)), persons must also comply with all applicable predicate rule requirements for record retention and availability such as (211.180(c) general requirements. The decision to follow 21 CFR part 11 should be justified and documented as part of a risk assessment and based on the value of the records over time. The FDA does not object to archiving of required records in electronic format to non-electronic media such as paper, or to a standard electronic file format (examples of such formats include, but are not limited to, PDF, XML, or SGML). Persons must still comply with all predicate rule requirements, and the records themselves and any copies of the required records should preserve their content and meaning. As long as predicate rule requirements are fully satisfied and the content and meaning of the records are preserved and archived, you can delete the electronic version of the records. In addition, paper and electronic record and signature components can coexist as long as predicate rule requirements are met and the content and meaning of those records are preserved.

Electronic Signatures

Electronic signatures are computer-generated character strings that count as the legal equivalent of a handwritten signature. The regulations for the use of electronic signatures are set out in 21 CFR Part 11 of the FDA. Each electronic signature must be assigned uniquely to one person and must not be used by any other person. It must be possible to confirm to the authorities that an electronic signature represents the legal equivalent of a handwritten signature. Electronic signatures can be biometrically based or the system can be set up without biometric features.

Conventional Electronic Signatures

If electronic signatures are used that are not based on biometrics, they must be created so that persons executing signatures must identify themselves using at least two identifying components. This also applies in all cases in which a chip card replaces one of the two identification components. These identifying components, can, for example consist of a user identifier and a password. The identification components must be assigned uniquely and must only be used by the actual owner of the signature.

When owners of signatures want to use their electronic signatures, they must identify themselves by means of at least two identification components. The exception to this rule is when the owner executes several electronic signatures during one uninterrupted session. In this case, persons executing signatures need to identify themselves with both identification components only when applying the first signature. For the second and subsequent signatures, one unique identification component (password) is then adequate identification.

Audit Trail

Title 21 CFR details predicate rule requirements relating to documentation of, for example, date time, or the sequencing of events, as well as any requirements for ensuring that changes to records do not obscure previous entries. Making the decision on whether to apply audit trails, or other appropriate measures, or on the need to comply with predicate rule requirements should involve a justified and documented risk assessment. Any risk assessment should determine the potential effect on product quality and safety and the integrity of the record.

Change Management

Validation programmes are subject to change control. Each company or organisation should have a procedure detailing the change management process.

Any system, facility, document or process that has the potential to impact product quality and the validated state is generally subject to following a change control process. Another term used in industry is enterprise change control or engineering change control. Essentially, these terms are the same. The intent is to control and manage change consistently.

A change control can take the form of a document which drives the agenda and the specific requirement. Change control is also created with enterprise software such as Kintana, Documentum and SAP. While each company will have varying processes, some basics are common. These include the three stages of change control; pre-implementation, implementation and post implementation (if required).

Validation Deliverables

The deliverables of validation activities should be in accordance with a project validation plan of validation master plan. For small projects or changes to computerised systems, a change control may serve as the validation plan. However, some typical deliverables include the following:

- GxP assessment (note, some systems may be non GxP applicable)
- User requirements specification
- Third party audit
- Validation plan
- Design specification such as functional, software, hardware and technical specifications
- GxP risk assessment
- Validation protocols
- Traceability matrix
- Validation report

8.0 FACILITIES, UTILITIES AND CLEANROOMS

8.1 Introduction to Commissioning, Qualification And Validation

The Qualification of facilities and utilities is best managed with the creation of a qualification plan. The plan can provide a framework that outlines the qualification activities, rationales, deliverables, resources and timing. However, certain qualification activities are strongly recommended and mandated by health regulators especially within pharmaceutical biotech, MedTech and medical device sectors. The regulatory legislation pertaining to the specific products and markets can inform the essential qualification requirements. Medical devices range in their principal mechanism of action, complexity and intended use. For example, the facility and supporting utilities necessary for the manufacture and packing of a surgical implant differs from an Orthopedic crutch or aid. Yet again, a medicinal product or combination device such as a pre-filled syringe with a biological formulation will require aseptic techniques to be applied during the process. This controlled environment that assures sterility is supported by qualified facilities and utilities that need to function and perform consistently. Therefore, the scope and complexity of C&Q and validations must be designed based on the products manufactured and their intended purposes. With that said, there are a number of keystone commissioning, qualification and validation activities that represent best practices that are broadly applied to meet regulations. The essentials of C&Q can be specified in company (in-house) procedures or standard operating procedures. The discrete requirements required for specific projects can then be guided with the creation of a C&Q plan.

High Level understanding C&Q

Requirement

Specification

Verification

Release

The above diagram introduces some key concepts that are applied during C&Q and validations. The intial stage deals with the requirements. Requirements should be specific, mesaable and unambigous. The requirements must decribe the use in my (intended use).

In response to requirements, a specification document is needed. The specifation document translates the user requirements into elements that help focus the contruction and design teams to meet the needs of the business. When contruction and assembly is completed the verificaiton activity can commence. Successful C&Q verification allows release of a system, factility or process for the subsequent validation activites and manufacturing.

8.2 REGULATORY REQUIREMENTS

If C&Q is to be applied within a pharmaceutical company, regulatory guidance in Europe is provided under Eudralex V4 Annex 15 Qualification and Validation.[1] The main stages include (i) user requirements (specification), (ii) Design Qualification (iii) Commissioning (iv) Installation Qualification, (v) Operational Qualification and (vi) Performance Qualification. The FDA and other regulatory bodies throughout the globe may require specific requirements to be considered for certain industries.

C&Q Model

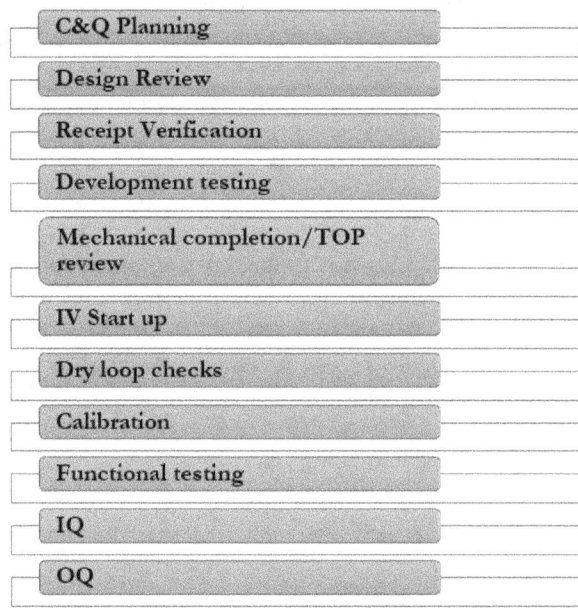

The above model is an example of a common approach to C&Q. The applicable model should be created by your company as the necessary product knowledge, quality, corporate and regulatory requirements are understood best by on which elements need apply.

1 Eudralex Volume 4, EU Guidelines for Good Manufacturing Practice for Medicinal Products for Human and Veterinary Use Annex 15: Qualification and Validation

8.3 Qualification Model for Manufacturing systems and Equipment

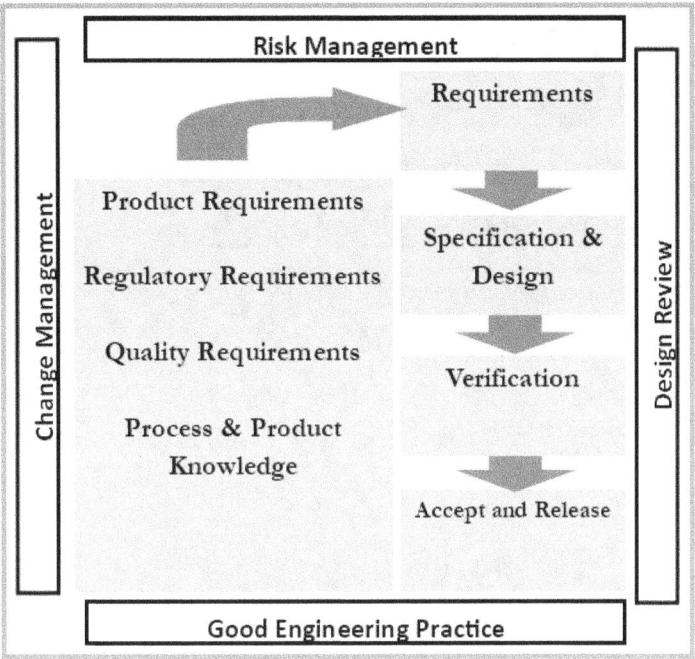

Further guidance on manufacturing systems and equipment in a pharmaceutical and biopharmaceutical context is issued by ASTM[2]

User requirements specification (URS)

A URS is a specification document that is written for equipment, facilities, utilities or systems that defines the specific requirements to meet the use application and intended use. This includes the physical, function, operational, electrical, performance requirements and so on.

Creating a written URS ensures that the user requirements are documented and approved and this can be the basis for design qualification and providing a vendor or OEMs, Original equipment manufacturers with a 'build specification' The URS is an important input during the qualification and validation process and therefore should include all critical and quality related requirements.

Design qualification (DQ)

The next element in the qualification of equipment, facilities, utilities, or systems is Design Qualification, DQ where the compliance of the design with GMP should be demonstrated and documented. The requirements of the user requirements specification should be verified during the design qualification process. The DQ should be approved by relevant roles.

---

2 ASTM, Standard Guide for Specification, Design and Verification of Pharmaceutical and Biopharmaceutical Manufacturing systems and Equipment

Commissioning

Commissioning includes the application of good engineering practices to introduce new equipment and facilities/utilizes into operation in a controlled manner that supports project delivery, safety and success.

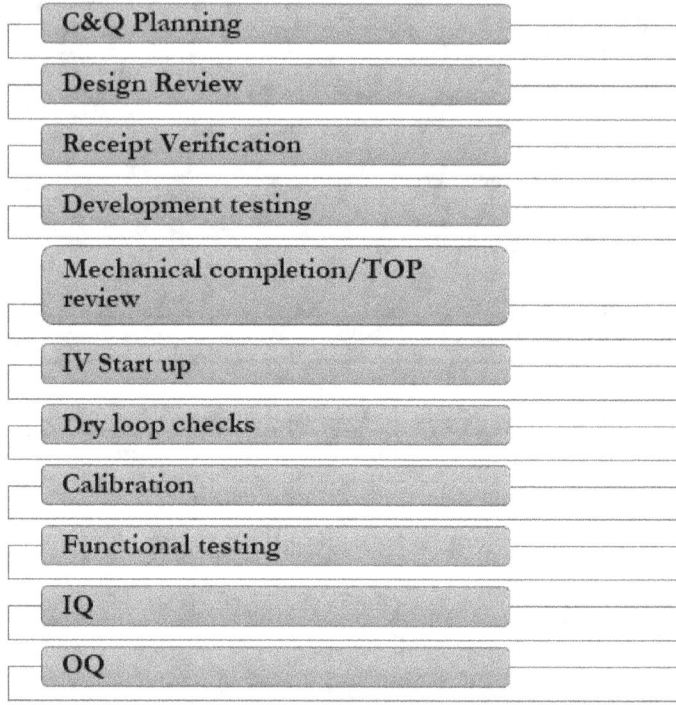

C&Q Planning

Design Review

Receipt Verification

Development testing

Mechanical completion/TOP review

IV Start up

Dry loop checks

Calibration

Functional testing

IQ

OQ

Commissioning is essentially a managed engineering approach to start up and provide turnover of equipment, utilities, systems and facilities. It involves field verification and review of system specific components and review of the construction, building and assembly of systems to ensure they meet the intended use and design specification.

After the commissioning stage, these systems are then turned over to the responsible person or owner which can then proceed with qualification and validation as required. Therefore, the real value of an effective commissioning program is reducing risks in qualification and providing the basis of success.

C&Q Planning

C&Q planning is achieved by creating a C&Q plan which sets out the activites, deliverables and key information of a project. It should cover all of the elements listed and provide guidance to the C&Q team.

It should maintained as accurate and updated if required periodically. Approval of the plan should include all stakeholders which provides a mechanism for agreeing the C&Q strategy and communicates the nuiances for each stage.

Design Review

It is best practice to include a design review of systems/facilities that are been introduced. The design review involves reviewing the proposed design (drawings, specifications, materials, configurations) against pre-appproved requirements such as the URS and other procedures (company SOPs) and standards that may need to apply.

For a direct impact system a design review is normally completed under the term Design Qualification, DQ. Design Qualification should include a review of the following documents:

- o List of the approved documents in scope of the review
- o Scope of DQ/review
- o Attendee list and function represented
- o List of open items
- o List of corrective actions
- o Conclusions that Design is suitable for intended use and that the project may proceed to the next stage.

Receipt Verification

Receipt Verification (RV) ensures that components or systems and assocaited ancillary items and documentation is provided and received as purchased per vendor P.O.s. RV is performed prior to installation, assembly and verification activities. Therefore is an incorrect system or equipment is provided it can be returned to the provider with minimum delay. RV checks include:

- o Confirmation of model number
- o Inspection to include equipment is not damaged
- o All specified items/components on P.O are fullfiled and reflected in delivery documentation.
  - o Documentaition and manuals are available.

Development testing

Development testing is the intial testing on a system that confirms that the system functions according to the design intent and the functions specifications. It can be completed in a simulated manner often termed 'off-line' testing. Common activities during this stage include I/O, (Input/ output)testing. Development testing, if documented accordingly following GEP and GDP standards and configuration controls may be leveraged to meet the requirements of functional testing or OQ, operational qualification testing

Mechanical completion/TOP review

Mechanical completion is confirmation that the system or equipment is phycically assembled and fixed in its use configuration. For stick built systems, drawings can be used to verify that the system is as designed and intended. Mechanical completion can also include construction activities. The contractor responsible for contruction and assembly signifies that the system is ready to handover to the C&Q team after Mechanical completion. The package of documentation is referred to as a turn-over package or TOP. Mechanical verification can occur prior to Installation verification (IV) as some checks and inspections may need access to restricted areas or areas subject to further modification. (e.g. pipr verifications prior to insulation). Open items or issues where the mechanical completion does not meet requiremetns or drawing specifications can be fixed in real time if feasible or alternatively may be tracked via a punch list. A punch list is a tracking mechanism where follow up actions, improvements or remediation is required prior to formal closure of the TOP package.

| IV Start up |
| Dry loop checks |
| Calibration |

Start up is the next step after installation verification is completed for each system. The safety of the system and its operation is important to verify, so start up is done in a controlled and safe manner. A start up protocol is best practice.

| Functional testing |

The purpose of Functional testing (FT) is to verify the safety of components during operation. The activity can also be referred to as functional verification. It requires a combination of physical inspections and also completed tests in person by testing alarms, valves, interlocks, emergency stops and so on. FT of a cleanroom during the commissioning stage would include temperature and humidity monitoring, particulate monitoring and differential pressure monitoring.

| IQ |
| OQ |

Installation qualification (IQ)

For the pharmaceutical, biotech and medtech industries compliance to regulations of health authorities, competent authorities and notified bodies is required in order to manufacture and produce products. The regulators grant marketing authorizations which require companies to be subject to audit and inspection. Validation of processes, equipment, facilities and utilizes is mandated by various regulations, depending on the markets and geography. US Federal drug association require validation for both medical devices, pharmaceuticals, biotechnology, therapeutics and combination products. Installation Qualification, IQ should be performed on equipment, facilities, utilities, and systems, including computerized systems.

In respect of manufacturing facilities, cleanrooms which are built-for-purpose environments designed to control particulate and climatic conditions. The type of equipment, facility or utility should inform the requirements of IQ. However, at a minimum the following should be considered:

o Verification of the correct installation e.g. pipe work and services as required by drawings and design specifications.
o Verification of the correct installation according to the supplier specifications and the intended use and purpose of the system.
o Calibration of instrumentation
o Verification of the materials of construction.
o Operating manual and supplier documentation
o maintenance requirements
o

Operational qualification (OQ)

OQ verification testing should include but is not limited to the following checks:
o Tests that have been developed from the knowledge of processes, systems and equipment to ensure the system is operating as designed
o Tests to confirm upper and lower operating limits, and /or "worst case" conditions

Completion of a successful OQ should allow the finalisation of standard operating and cleaning procedures, operator training and preventative maintenance requirements.

Performance qualification (PQ)

PQ should normally follow the successful completion of IQ and OQ.
PQ should include, but is not limited to the following:
o Tests, using production materials, qualified substitutes or simulated product proven to have equivalent behavior under normal operating conditions with worst case batch sizes.

The frequency of sampling used to confirm process control should be justified; ii. Tests should cover the operating range of the intended process, unless documented evidence from the development phases confirming the operational ranges is available for manufacturing.

Re-Qualification

Equipment, facilities, utilities and systems should be evaluated at an appropriate frequency to confirm that they remain in a state of control.

re-qualification is necessary and performed at a specific time period, the period should be justified and the criteria for evaluation defined. Furthermore, the possibility of small changes over time should be assessed.

8.4 Cleanrooms

A specially constructed enclosed area has the following controlled parameters:
- o Temperature
- o Humidity (Relative Humidity)
- o Sound and Vibration
- o Lighting
- o Airflow Pattern
- o Pressurization
- o Particle Count
- o Microbial Contamination
- o Gaseous Contamination

Heating, Ventilation and Air-Conditioning (HVAC) contributes to the functioning of clean zones. It works to prevent any negative effect on production due to changes in climatic conditions. In addition, it also works to prevent product contamination and providing ergonomic working conditions.

Good Engineering Practices, application of standards, regulations and commissioning and qualification planning are necessary to deliver systems that are fit for purpose and perform as required.

- International Organization for Standardization (ISO), ISO 29463 - High-efficiency filters and filter media for removing particles in air, Parts 1 to 5.
- International Society for Pharmaceutical Engineering (ISPE) – Good Practice Guide – Heating, Ventilation and Air Conditioning (HVAC)
- International Organization for Standardization (ISO), ISO 14644 - Cleanrooms and associated controlled environments, Parts 1 to 9.

- FDA 21 CFR Parts 210 and 211 – Current Good Manufacturing Practice In Manufacturing, Processing, Packing or Holding of Drugs; General and Current Good Manufacturing Practice For Finished Pharmaceuticals
- PDA Technical Report No.13- "Fundamentals of a Microbiological Environmental Monitoring Program"
- EudraLex Volume 4, EU Guidelines for Good Manufacturing Practice for Medicinal Products for Human and Veterinary Use, Part 1, Chapter 3: Premises and Equipment)
- EN 1822 Series "High efficiency air filters (HEPA and ULPA)"
- EN 779 "Particulate air filters for general ventilation. Determination of the filtration performance."
- EN 1886 "Ventilation for buildings – Air Handling Units -Mechanical Performance"
- EN 12464-1 – "Light & Lighting of Indoor Work Places".
- ASHRAE Handbooks – Fundamentals, HVAC Systems and Equipment, HVAC Applications, Refrigeration
- ASHRAE Standard 110 – "Method of Testing Performance of Laboratory Fume Hoods"

- ASHRAE 52.2-1999 "Method of Testing General Ventilation Air-Cleaning Devices for Removal Efficiency by Particle Size"

The environment where products are manufactured, processed and packaging can lead to contamination issue that may impact the product and safety. Therefore, an appropriate environmental cleanliness level is required to minimize the risks of particulate or microbial contamination to the product. The levels of cleanliness depends on the activity and products been provided. A cleanroom is defined as enclosed area which is environmentally controlled with respect to particles, temperature, humidity, air pressure, air pressure flow patterns, air motion, vibration, noise, viable organisms, and lighting and is designed and constructed for the intended use in mind.

ISO 14644-1[3] defines a cleanroom as *"a room in which the concentration of airborne particles is controlled, and which is designed, constructed and operated in a manner to control the introduction generation, and retention of particles inside a room".*

ISO 14644-4 A.1[4] suggests clean rooms are *"enclosed (rooms)or surrounded by further zones of lower cleanliness classification. This can allow the zones with the highest cleanliness demands to be reduced to the minimum size. Movement of material and personnel between adjacent clean zones gives rise to the risk of contamination transfer, therefore special attention should be paid to the detailed layout and management of material and personnel flow"*

Critical process areas are more stringently controlled portion of the cleanroom. Pharmaceutical cleanrooms and controlled zones should;

- Prevent the quality of products being impacted with unwanted

airborne contaminants or particles and prevent products from contaminating each other

- Provide a comfortable environment for the operators and limit

exposure to hazardous risks (e.g. drug particulates, fumes, vapors)

- Remove any contaminants form the room as effectively as

Possible and in accordance with regulatory requirements.

Cleanroom Zoning and Classification

Selecting a suitable classification for a room or manufacturing facility depends on several factors. Firstly, it can be said that sterile products require a more stringent set of criteria than non-sterile products. However, there is an extensive range of products and medical devices that are sterile but are used in different ways and consist of different materials and technology. Some sterile products are single use only and used for short-term purposes and then disposed of.

Other sterile products are used subcutaneously for longer periods or even require implantation. Therefore, the design of a facility along with its HVAC specification must be appropriate to the product being manufactured. High-risk products require greater control.

---

[3] ISO 14644-1 Cleanrooms and associated controlled environments Part 1: Classification of air cleanliness by particle concentration.

[4] ISO 14644-4 Cleanrooms and associated controlled environments — Part 4: Design, construction and start-up.

The goal of facilities and HVAC systems is to minimise contamination and the associated risks. Using a sterile versus non-sterile rule of thumb is not adequate when classifying a room or facility.

Standards including EN ISO 14644-1 and guidelines such as EU cGMP Guidelines EudraLex volume 4 Annex 1 (2008) should be consulted in order to fully understand the requirements of each ISO classification and grade of room.

ISO classifications do not specify room occupancy states but when a designation is applied, the occupancy state must be stated in the relevant documentation or procedure. The most relevant European Guideline (Annex 1 of the EU cGMP Guideline) lists four classification grades and their associated particulate limits in the 'at rest' and 'in operation' conditions. In general, for the sterile and non-sterile products, similar classes are applied, but in non-sterile production the producer could assign their classes, having similar particulate concentration, temperature, pressure etc. but lower air-change rate could be used.

The classification of a cleanroom is determined by the maximum number of particles acceptable according to a specific size and per the volume of air. The selection of the right classification for any given cleanroom needs to consider the application, the type of products been processed and the type of processes. For example, a product that can be terminally sterilized generally requires less control and can be sealed in a area that is not fully aseptic.

Particles in the air is made up of both Viable and Non-Viable Particles. Viable particles can present microbial risks. The levels of viable and non-viable particles is an indication of how 'clean' an area is. Therefore, monitoring these levels is useful in determining any adverse trends and tracking on an ongoing basis the cleanroom is operating as required.

Critical areas such as ISO Level 5 are afforded protection by areas of a lower classification. Raw materials, components and personnel are controlled with an increasing level of cleanliness in order to prevent contamination from the outside impacting the critical zones.

Types of Contamination
o   cross contamination (of a product/material with another product/material)
o   non-microbial particulate contamination (non-viable particles)
o   biological/microbiological contamination (viable particles/micro-organisms)
o   Factors Influencing Contamination Cleanliness Levels in the Manufacturing Processes:
o   process
o   air cleanliness
o   personnel hygiene and clothing
o   work practices
o   material design (material of construction, surface finishes, room finishes, equipment, open system/enclosed system, utensils etc.)
o   material cleanliness

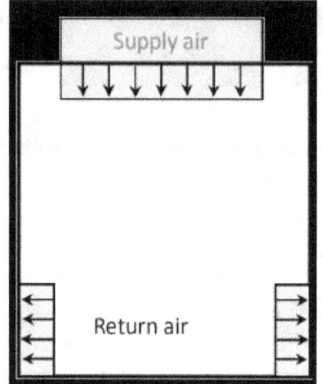

As built is the condition where the installation is complete with all services connected and functioning but with no production equipment, materials or personnel present.

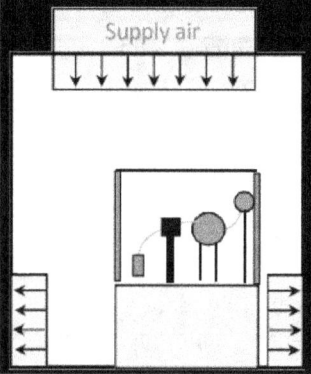

At rest condition is where the installation is complete with equipment installed and operating in a manner agreed upon by the customer and supplier, however, no personnel present.

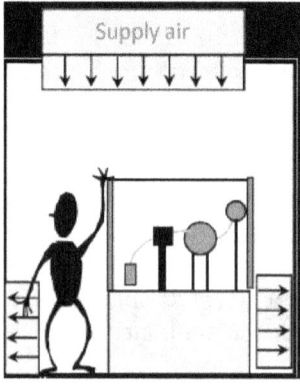

In operation is the condition where the installation is complete with equipment installed and operating in a manner agreed upon the customer and supplier and where personnel present and working.

ZONE CLASSIFICATION

Applying ISO 14644-1 rules, cleanrooms are classified based on the level of airborne particulates within the environment. ISO Class 5-9 are summarized below, with ISO Class 9 allowing the greatest levels of particulate.

ISO ZONE 5

Critical zones are areas or clean room zones where the product, packaging, or closures are exposed to environmental conditions during the completion of the last manufacture steps. This exposure to the environment may impact product quality. This control of the critical zone is achieved by the design of the room, use of HEPA filtration during HVAV, gowning requirements, access control and the control and monitoring of conditions such as temperature, relative humidity and pressure differentials.

ISO Class 5 permits a maximum allowable particles per cubic meter: 3,520
Airborne particulates are limited to a very low level, making it suitable for environments requiring extremely high levels of cleanliness, for example, pharmaceutical production and Aseptic manufacturing and biotech.

ISO Zone 6

ISO Class 6 permits a maximum allowable particles per cubic meter of 35,200. Therefore, this cleanroom has a higher particle count compared to ISO 5 cleanrooms but are still maintained to high cleanliness standards.

ISO Zone 7

The maximum allowable particles per cubic meter is 352,000
ISO Class 7 cleanrooms have higher particle counts compared to Class 6

ISO Zone 8

ISO Class 8 permits a maximum allowable particles per cubic meter of 3,520,000. ISO Class 8 cleanrooms have even higher particle counts compared to Class 7 and are considered as controlled environments with a lower degree of cleanliness E.G. food processing, or certain medical device manufacturing processes.

ISO Zone 9

ISO Class 9 permits the maximum allowable particles per cubic meter of 35,200,000. ISO Class 9 cleanrooms are used when high levels of cleanliness is not critical.

ISO classifications ensure that cleanrooms function at the appropriate levels of cleanliness to support the operations and activities completed within them.

HVAC Particulate Control

The main purpose of the HVAC system in a cleanroom is to ensure the processing environment does not negatively impact upon product quality. Prior to the design and specification of a HVAC system, the product(s) and processes need to be understood to assess and determine the environmental controls necessary for a particular product, taking into account the type of product, the product specification and packaging and the regulatory requirements of competent authorities and notified bodies.

Total Airflow Volumes & Recovery Rates

Air change rates per hour, (AC/hr) are an important factor in contamination control. The arbitrary 20 AC/hr are a result of previous industry standards, however, nowadays, the number of air changes and depends on several factors including:

- Particle Generation Rate, (PGR) inside the space from people, equipment, etc.
- Room supply air volume
- Quality of air distribution (ventilation efficiency)
- Cleanliness of dilution air (negligible if HEPA filter are used)

Particle Generation Rate (PGR)

PGR is a measure of the number of viable and non-viable particles being generated in the cleanroom from both people and equipment and to a lesser extent the building fabric as it should be designed to be non-shedding. Good cleanroom gowning and personnel training are an important factor in reducing room particle levels and AC/hr.

Room Supply Air Volume

The supply air volumetric flowrate to a room is not only determined by a required particle level in the room but also by several other interrelated factors;

- o Room heat gains (internal and external)
- o Number of occupants in the space and activities
- o Gowning levels
- o Moisture gain to the space from internal and external influences
- o Room leakage and differential pressure requirements

Heat and humidity gain are typically more easily controlled but should be considered by the HVAC designer.

Particle generation and removal is generally the main driver of the supply air volume and hence air change rates in cleanrooms.

HVAC designers default to "rules of thumb" for supply air rates by class of space, rather than calculating the actual airflow rate based on the activities in the room.

Non-unidirectional flow & unidirectional

Non-unidirectional flow uses air turbulence and dilution to mix particle contamination generated by people and machinery in the clean room. Clean Filtered air is delivered to the room through ceiling mounted air diffusers. This air mixes with the room air and removes airborne contamination through air extracts generally at low level in the walls.

For large rooms swirl diffusers induce room air vertically up to the diffuser to mix with the supply air. These diffusers create good dilution of contaminants in the room over the perforated diffuser type and may be used in rooms where there is minimum dust liberation.

However, they should be avoided in rooms where excessive dust is generated as they would add to the distribution of the dust and could be hazardous for the operators.

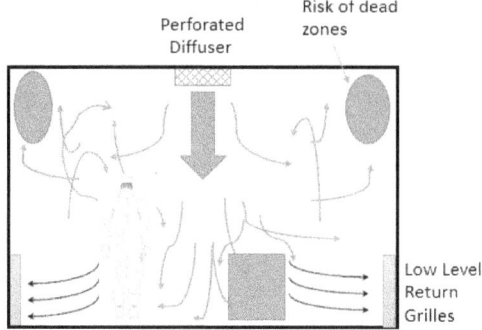

Non-Unidirectional flow with Perforated Diffuser

A shortcoming of non-unidirectional cleanrooms is the creation of air dead zones with high particle counts.

These pockets can persist for a period of time, and then disappear. This is due to currents that are set up within the room due to process related activity combined with the random nature of the downward airflow. Airflows should be planned in conjunction with operator locations, to minimize contamination of the product by the operator.

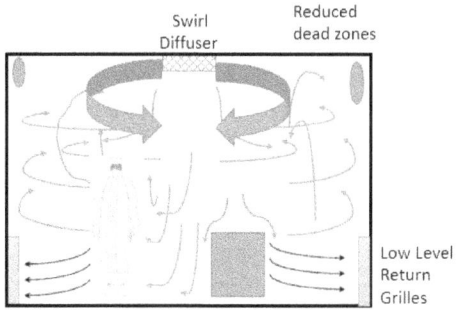

Swirl diffuser

Unidirectional flow

Unidirectional airflow (aka not laminar) is defined by ISO14644-1 as a "controlled airflow through the entire cross-section of a cleanroom or a clean zone with a constant average velocity steady velocity and air streams that are considered to be parallel".

Unidirectional airflow is achieved by supplying filtered air through 95-100% of the ceiling. The air moves vertically downward laterally from the ceiling to return air grilles at low level.

This approach allows the contamination generated by the process or surroundings to drift to the floor level where they are extracted. This is known as the displacement method as it develops minimum air turbulence.

Unidirectional airflow velocity should be uniform and sufficient to dilute and remove particles generated in the room before they settle on a surface.

The particles are finally captured by the low-level return grilles and returned through the return walls and recirculated through the filters in the AHUs or ceiling Fan Filter Units (FFU). Cleanrooms with classification rating Zone 5 or below are almost invariably designed for unidirectional airflow.

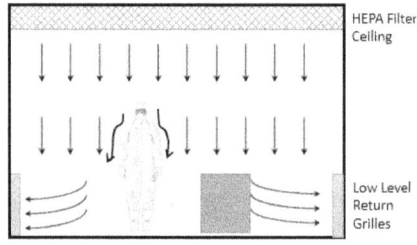

Unidirectional (turbulent or dilution) Airflow

Airlocks

Airlocks are small rooms (typically anterooms or material transfer) with interlocked doors, constructed to maintain airflow gradient and air pressure control between adjoining rooms (generally with different air cleanliness standards).

The primary role of an airlock is to provide a pressure buffer between areas of different classification, and for that reason its own internal pressure is somewhere between (floats), or equal to, the rooms it connects.

If a door is opened, the adjoining rooms will be at the same pressure, and contamination can flow across the doorway, most easily moved by personnel or equipment dragging the contamination along.

Therefore, once the doors are closed, there must be enough air changes to reduce the contamination level before the next door opens (this is one of the few places where air changes are important (quick room recovery).

So, the higher the air change rate in the airlock, the faster the recovery and the less chance of contamination passing into the next room.

Recovery to zero counts really can't happen with personnel in the room, but the airlock presents a relatively SMALL volume of lightly contaminated air that may be dragged through the next door to be opened.

Airlocks are generally divided into two categories: Personal Airlocks (PAL) and Materials Airlocks (MAL), these control personal and material flow into and out of clean spaces through a series of interlocking doors. Airlocks also help to maintain room pressure differentials between rooms of different classifications.

There are typically three (3 No.) types of airlock pressure arrangements used:

1. Cascade - airflow from areas of higher pressure, through the airlock to the area of lower pressure.
2. Bubble – airlock is at highest pressure to surrounding rooms, air flows from the airlock to the clean rooms.
3. Sink – airlock is at lowest pressures to surrounding rooms, air flows from the clean room and corridors.

Sink or Bubble configuration are only used to restrict cross contamination of products between rooms. Cascade is the typical arrangement used where the airlock pressure floats between the pressure in both rooms.

Door interlocks: An interlock system or visual or audible alarm system is recommended to prevent the opening of more than one of the airlock doors at a time. Interlocked doors prevent an air flow through the airlock. Interlocks should be disarmed in case of emergency.

Pass-throughs: Small material airlock, called pass-throughs (PT) see Photo 7.2, which are too small for personnel, are used to transfer product from the higher-class rooms to lower class areas. Pass-throughs usually have interlocking sliding doors to also maintain clean room pressures between two different zones.

PT fall into two categories, namely dynamic or passive. Passive PT is typically used in cleanrooms with interlocked doors. PT shall be big enough to receive small items for example batch cards, samples, small consumables. Room Differential Pressure **(ΔP)**

As most facilities consist of multiple rooms with different requirements for cleanliness, differential pressure is required between the cleanrooms to ensure airflow from the cleanest spaces to the least clean spaces.

Certain operational activities may require a pressure differential to be maintained between rooms with the same classification but require an air pressure cascade (e.g. Autoclave Loading vs. Unloading). Where required this pressure must be less than the difference between room of different classification, typically +5 Pa. The requirement is to be determined at design stage and verified at certification.

The pressure differential of a cascade airlock is measured across the airlock and not across each door. Therefore, when only one door of an airlock is opened, a measurable DP between the cleanrooms persists. It also ensures the room pressure of the highest cleanliness room is maintained at a reasonable level.

Room Construction: Hard-ceiling construction is preferred for pressure-controlled spaces. In addition, air migration above the ceiling should be minimized between controlled and uncontrolled spaces.

To maintain pressurization within rooms, all doors should be fitted with continuous seals, manufactured of materials acceptable for cleanroom operation and wipe down. The gap between the finished floor and the bottom of door should be uniform at approximately 5mm (3/16 inch) when closed.

Door floor sweeps are not recommended for swing doors due to their accumulation of dirt, scratching of floor, and maintenance. Doors preferably should operate, such that the pressure differential pushes the doors closed against the frame.

Should the doors open to the low-pressure side, the door closer springs should be sufficient to hold the door closed and prevent the pressure differential from pushing the door open resulting in excessive leakage.

Air Leakage: Care shall be incorporated during design and construction to eliminate air ex-filtration within classified and controlled spaces to reduce air make-up and to maintain better static pressure control.   The following, which is non-exhaustive, identifies areas where leak occurrences are most prevalent:

- Ductwork & Pipework penetrations through walls and termination into rooms
- Door perimeter
- Door closure mechanisms
- Surface Interfaces
- Spaces around equipment
- Access panels
- Electrical Fittings
- Light fixtures

The airflow leakage rate should be calculated for each room. This calculation should be based on the known architecture and the design pressure differential established for the facility

Differential Pressure Monitoring: For classified cleanrooms, it is recommended that pressure differential between cleanrooms be monitored and recorded continuously throughout each shift. There are 2 methods of measurement commonly used to monitor room pressure:

- Room to Room – Differential room Pressure
- Room to common reference point

Room to Room measurement directly meets the GMP requirement for room DP monitoring which clearly indicates the pressure relative to an adjoining room. This method is preferred for monitoring only i.e. by an EMS. When automatic room pressure control is used (VAV), the preferred measurement is room to a common reference point for stability.

There is no GMP requirement that room pressure or airflow is automatically controlled but it is recommended in clean room areas to ensure pressures are maintained. A BMS is used for automatic control and monitoring and non-critical alarming.

Fixed/manual damper control (CAV) on the supply air or return air is not recommended for cleanrooms due to its inability to compensate for fluctuations in supply and return fan output, filter loading, and exhaust modulations.

Airflow variations from dust collecting, vacuums, or process systems, and their effect on space pressurization, should be accounted for in the operation of the HVAC system.

Room Temperature and Relative Humidity

The ratio of the actual water vapour pressure of the air to the saturated water vapour pressure of the air at the same temperature expressed as a percentage.

More simply put, it is the ratio of the mass of moisture in the air, relative to the mass at 100% moisture saturation, at a given temperature.

The normal operating temperature requirement for each classification .Temperature and humidity must be appropriate to the product and process. Consideration should be made for specific product and process requirements.

A facility should meet stated relative humidity design conditions; however, the acceptability of the facility or operation depends on meeting the operating ranges.

Room Temperature and Relative Humidity (RH) requirements depend on the product requirements and operator comfort. A risk assessment should be completed to determine the criticality of temperature and humidity on product quality.

Room temperature and RH and determined both are not critical parameters as the majority of product are elastomers whose impact temperatures have a wide band. Therefore, Temperature and Humidity should only be monitored and controlled for human comfort and kept within a range to ensure human discomfort (e.g. perspiring or dehydration) does not indirectly impact product quality.

As Temperature and RH are considered non-critical parameters they should be monitored and controlled on the non-validated Building Management System (BMS). If a Risk Assessment determines that Temperature or RH are critical parameters that may impact product quality, then these parameters should be monitored by a validated Environmental Monitoring System (EMS).

It is important to define the distinction between the design and operating parameters of a space. In Figure 7.10 below, Values of Critical Parameters of a Product indicates the relationship between the design, normal operating, and qualified (validated) operating (product stability range) ranges.

Temperature

Room temperature is maintained for personnel comfort, considering working activities, and should take into consideration the various gowning levels worn in each area to ensure the majority of personnel are comfortable.

Energy usage should also be considered when selecting room temperature verses levels of gowning. Most HVAC systems have a lower limit of 16°C & 75%RH without additional expensive HVAC equipment with higher running costs.

The temperature sensor should be specified with an accuracy ± 1°C (± 2°F) and be fully adjustable and calibrated annually.

Relative Humidity (RH)

Room Relative Humidity (RH) for personnel comfort should consider working activities to evaluate and determine the normal operating range and alert limits, while the product environmental requirements, if any, determine the qualified operating range where RH may have an impact on product. Where relative humidity levels are not specified as having product impact (e.g. aqueous product) the outer limits of the operating range or validation acceptance criteria shall be based on the following; In areas with a requirement for low particles and dust generation (e.g. clean room) the humidity should be maintained above 30% to minimize the risk of static electricity and particle generation due to dryness. Also, liquid products can lose moisture to a low-humidity space/room over an extended period. The risk of condensation and microbial growth increase above 70%RH. As a result, the Lower (Low-Low) and Upper (High-High) outer limits of the operating range or validation acceptance criteria (Alarm Limits) should be set at 30% and 70% in classified areas.

For Warehouse areas humidity levels should be maintained between 0-90%RH.

Control and Spread of Smoke

Systems shall not encourage the spread of smoke and fire, and in some instances, may be required to provide positive control. Careful attention must be paid to how smoke will be controlled and eliminated. It is important that smoke levels be quantified, with the necessary containment level established (i.e., cfm [m/s] smoke passage through the required smoke barrier), based on the type of structure in question, the characterization of the occupants, and their expected time to egress. Examples of positive control include pressurization of escape routes and smoke venting systems. All systems shall comply with local authorities' requirements at a minimum, and where applicable should follow the Uniform Building Code (UBC), the International Building Code (IBC), and the National Fire Protection Association (NFPA) Codes 45 and 101.

CLEANING

The selection of cleaning methods for cleanrooms and the sited equipment should be confirmed early in the design process because the selection may affect other design features (e.g. construction and finishes, cleanroom layouts, auxiliary services, etc.). Effectiveness of cleaning should be addressed in the validation, as applicable. Physical cleaning should be controlled by procedure and be recorded as specified.

8.5 HVAC Systems

Heating, ventilation and air-conditioning (HVAC) provide a critical function in the manufacturing of medical devices, pharmaceutical and biotech products by contributing to the quality and environmental conditions during manufacturing, processing and packaging. Temperature, relative humidity and ventilation should be appropriate and should not adversely affect the quality of pharmaceutical products during their manufacture and storage, or the accurate functioning of equipment.

Design parameters and user requirements should, therefore, be set realistically for each project, with a view to creating a cost-effective design, yet still complying with all regulatory standards and ensuring that product quality and safety are not compromised.

HVAC System Design

The HVAC system must be appropriately selected using the specific design requirements as outlined above. The system must be able to provide clean, conditioned air to the specified areas to meet all of the quality requirements. The most important precursor to HVAC design is the comprehensive definition of the function and performance required followed by the selection of an appropriate system. A poor selection can lead to unnecessarily high-energy consumption, and operational deficiencies. HVAC systems can be divided into two main types:

All-air systems rely on the movement of large quantities of air through a central air handling unit to control room conditions, as well as provide for ventilation requirements.

They have the advantage of being relatively simple with most of the unit situated in one location; however, they are very space consuming. All-air systems tend to be relatively inflexible and not ideal for areas that are likely to need environmental alteration on a regular basis.

These HVAC systems are used for areas that have a lot of small zones, each with slightly different thermal loads but which requires constant ventilation. These systems can have poor energy efficiency if a lot of reheat is required. These are typically used in large manufacturing areas, and laboratories with many small rooms.

HVAC systems are typically situated above production areas, though this isn't always the case. Air Handling Units (AHUs) are usually located on technical floors. Air is distributed through various channels:

- o   Above false ceilings

- o   Through shafts (on the surface!!)

- o   Through double-wall clean room walls

Design requirements shall be established in a User Requirement Specification (URS) document.

Cleanrooms may have more than one classification shared among adjoining suites, depending upon manufacturing, research and development, and containment requirements.

The cleanroom, or main assembly area, shall be specifically designated either as a specific ISO classification or Controlled Not Classified (CNC) classification, however, adjoining spaces may be designated an alternate class, and controlled via differential pressure requirements.

Other factors can affect the environmental conditions within the CR and/or CNC. Examples of these factors include the following: Number of personnel occupying each area Number and types of equipment Cleaning frequency (e.g. equipment and facility) Personnel gowning Airflow (e.g. directional, turbulent, and rate per hour) Training (e.g. movement, behavior, hygiene) Factors, such as these, may affect the cleanroom system and should be considered in the design criteria and prescribed in the user requirement specification document(s).

A well designed environment is constructed with materials that allow for ease of cleaning and sanitization. Current Good Manufacturing Practices (cGMPs) require that buildings be of suitable size, construction, and location to facilitate cleaning, maintenance, and proper operation. Additionally, the cGMPs are concerned with the potential contamination and cross contamination of product.

Based on the environmental needs of the product and/or process, controlled environment cleanrooms and areas are designed to separate manufacturing operations, and minimize the potential for contamination. The category and level of contamination control required by the product will help determine the Abbott Vascular room categorization. Abbott Vascular room classifications are based on non-viable air particulate requirements during At Rest conditions. Table 1 reflects minimum environmental specifications by Room Classification.

HEPA filters and Dehumidification

For most HVAC applications, dehumidification is best achieved by the use of cooling coils. It should be noted that dehumidification is a very high consumer of energy and should only be used if there is a real process need. When areas are not in use, the dehumidifier should be turned off, if possible.

When room humidity must be maintained below 50% during warm weather, an absorption dryer may be necessary unless the room temperature can be increased within specification to compensate.

Normal practice is to optimise size and efficiency of the absorption dryer by first removing as much moisture from the air as possible by cooling. The design of absorption dryers is normally based on a slowly rotating desiccant wheel.

Air is passed through the wheel and dried by the desiccant coating (guidance: lithium chloride especially if the wheel is not used frequently and silica gel if used permanently and with low humidity target). It is not normally necessary to size a dryer to handle the entire air volume. Drying a proportion of air and re-mixing to achieve the desired moisture content is usually sufficient.

Air humidification may be necessary during cold weather when introducing fresh air to spaces that require humidity control. When air humidification is necessary, humidifiers should be selected on the following basis:

> ➤ direct steam injection using steam
> ➤ direct steam injection using self-generative electric or gas steam humidifier.

Humidifiers should be located before the fan and the final filter which will remove any particulate generated. At least 300 mm clearance should be allowed upstream and 1 m downstream between humidifier manifolds and coils, attenuators etc. (general recommendation to be confirmed through calculation note provided by the vendor). A single manifold or multiple manifolds in parallel may be used to meet the humidification requirements as per manufacturer's recommendations.

Sound Attenuators

Sound attenuators should be provided as necessary, to achieve the specified noise levels within occupied spaces. To minimise external noise nuisance, assessment can confirm the necessity to use acoustic media (enveloped in polyester film), that is inert and corrosion-resistant at normal operating conditions. Material quality shall be equivalent to that specified for HVAC unit or ducts. Sound attenuators should be installed in the air handling unit or ductwork. The use of sound attenuators in the air supply and air return should be based on requirements for fresh air inlet and air exhaust, and according to external noise levels that might need to be maintained at or below the ambient site noise levels.

Dampers

The provision of sufficient dampers is essential for proper control. To minimise noise transmission into the room, these should be mounted as far as possible from the diffuser.

Carefully evaluate the space-by-space pressure control that will be used in the design. Static pressure control via hard balance or dynamic control via air terminal control units are both appropriate. Consideration should be given to the overall project size, the complexity of the facility and the project budget.

Automatic volume controllers are recommended for regulating air volume independently of supply pressure. They can be selected for constant volume, variable volume or dual duct mixing applications. Automatic low-leakage fresh air and exhaust air shutoff dampers are strongly recommended to isolate the HVAC network. Fresh air dampers shall be Class 3 minimum (maximum leakage preventing coil freezing). Whenever fumigation is performed shutoff damper shall ensure Class 4 leakage rate. Where dampers are required to provide modulating control of airflow, they must be selected to provide an appropriate level of control authority. This will normally mean a damper smaller than the duct size.

Heating and Cooling

Heating mode: Low pressure hot water (LPHW) is the preferred heating medium for HVAC applications and should be used whenever practicable. Electrical heating should be avoided due to fire risk and should be limited to low power coil and in locations where no other energies are available. Hazard operability analysis (HAZOP) must be conducted if electrical heating is being considered. Cooling mode: Chilled water is the preferred cooling medium for HVAC applications and should be used whenever practicable.

The direct expansion of refrigerant in coils is an acceptable method of cooling, particularly on small isolated plants, or where lower temperatures are needed for dehumidification or for cold room. This system, however, does not normally give close control. Direct expansion coils should only be used with extreme care on variable air volume systems (if speed driver available on compressors).

Heating Coils

The face velocity of air across heating coils should not exceed 2 m/s. Coils should be made of material suitable for applicable constraints. Drains shall be located outside the casing of the HVAC unit. Coils shall be removable.

Cooling Coils

Cooling coils have been identified as potential sources of microbial contamination; therefore, careful design is required to prevent water carryover and to ensure that drain pans do not retain water. Double tube, non-welded units are recommended. The face velocity of air across cooling coils should not exceed 2 m/s. Where necessary, stainless steel or plastic eliminator blades should be provided to prevent any moisture carryover. Where provided, these must be removable for cleaning.

Ductwork

For most applications, galvanised steel ductwork will be the most appropriate form of construction; however, stainless steel or plastic construction may be necessary where there is a higher risk of corrosion due to moisture or fumes (exhaust ducts usually). Where operating pressures above 2,000 Pa are necessary, fully welded construction is recommended. For contained ducts (e.g., exhaust duct before bag-in / bag-out filter), air tightness Class C shall be followed (EN 12237). For BSL-3, fully welded construction should be considered.

Generally ductwork should be constructed to an appropriate local standard, suitable for the maximum design pressure (positive or negative), such as those published by Sheet Metal and Air Conditioning Contractors' National Association (SMACNA) in the USA, Building and Engineering Services Association (B&ES) in the UK
Where flexible connections are proposed these must be designed for the same pressure as the ductwork. Solid ducted connections are preferred for final connections to terminal HEPA filter housings. For applications where flexible connections to diffusers are used, these should be no longer than 500 mm and nominally straight.

Special consideration must be given to fume extract ducts where these pass through fire barriers. Using fire dampers should be avoided where the loss of extraction could make a fire situation worse. An alternative design, such as the use of fire-rated ductwork, may be necessary in these cases. A thorough risk assessment must be conducted.

Temperature

Unless otherwise required by the regulatory, product and/or process driven specifications, HVAC system design should be based on user selected temperature(s) within the range defined in Table 2. A facility should meet stated temperature design conditions; however, the acceptability of the facility or operation depends on meeting the operating ranges.

Air Handling Units

All GMP Air-Handling Units (AHU) should have the capability of being custom designed and constructed to meet the more stringent operation and maintenance requirements for these areas. All GMP AHUs should be located in an internal plantroom to avoid risk of contamination during maintenance.

The frame shall be constructed from heavy gauge box section steel or robust aluminum structure and supported on a sectional steel channel. The AHU Casing shall be constructed of modular double skin panel which shall be of cold bridge free construction. Casing panels shall be manufactured from sheet steel, with galvanized inner skins and pre-painted outer skins. Panels shall be 60mm thick and filled with CFC free water-based PU insulation foam and shall be FM approved and meet NFPA fire rating. Casing shall have a U value of less than 0.55 W/m^{2}K. Enclosure panels shall be manufactured from 1.3 mm galvanized sheet steel and painted PVC coated finish.

All surfaces exposed to airstream shall be hot dipped galvanized or Type 304 stainless steel as indicated on the datasheets. Aluminum will be accepted in lieu of galvanized or stainless steel

An AHU with a cooling coil shall have a drift eliminator and stainless-steel drain trays of adequate size to collect water with a minimum depth of 40mm. The drain trays shall be extended past the coils to capture all moisture carryover from the coil by the airstream. All trays shall be inclined towards the drain connection and traps adequately sized to resist the positive or negative pressure at that point in the AHU.

Filtration

Air filters are the primary method to reduce contamination levels in an air stream and play a very important role achieving the clean room environment. It is not only the efficiency of the filter that is important to address, but also the energy consumption (the pressure drop during the entire operation).

The air change rates should be determined by the manufacturer and designer, taking into account the various critical parameters using a risk based approach with due consideration of capital and running costs and energy usage. Primarily the air change rate should be set to a level that will achieve the required clean area condition.

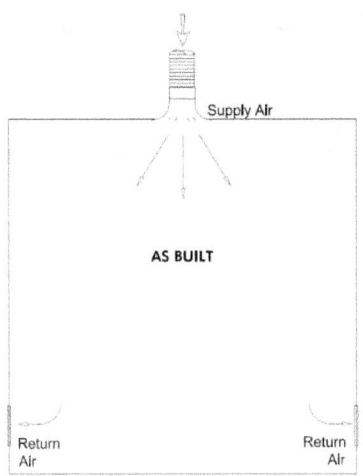

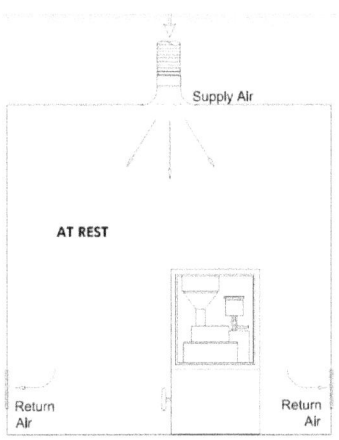

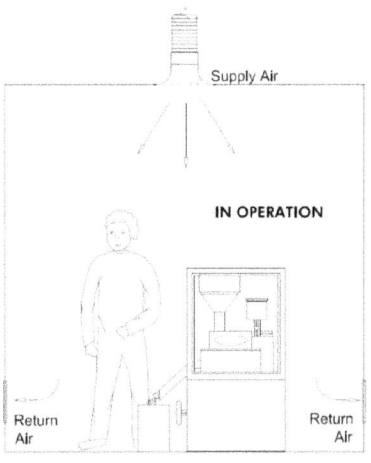

Room classification tests in the "as-built" condition should be carried out on the bare room, in the absence of any equipment or personnel. 4.1.9 Room classification tests in the "at-rest" condition should be carried out with the equipment operating where relevant, but without any operators. Because of the amounts of dust usually generated in a solid dosage facility most clean area classifications are rated for the "at-rest" condition.

Room classification tests in the "operational" condition should be carried out during the normal production process with equipment operating, and the normal number of personnel present in the room. Generally a room that is tested for an "operational" condition should be able to be cleaned up to the "at-rest" clean area classification after a short clean-up time. The clean-up time should be determined through validation and is generally of the order of 20 minutes.

Materials and products should be protected from contamination and cross contamination during all stages of manufacture for cross contamination control.

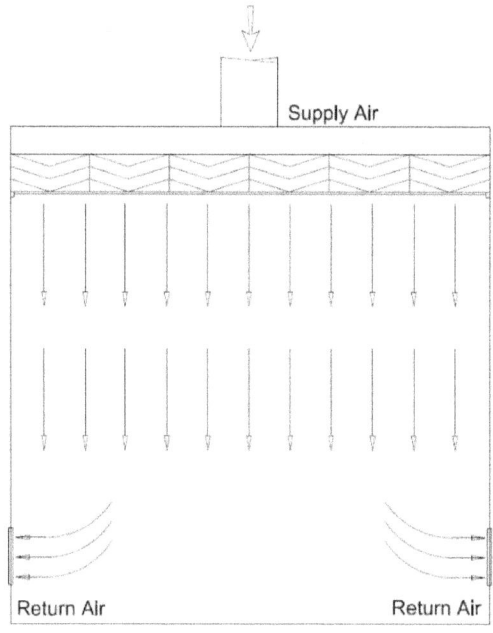

Course/ PRE- Filtration

Pre/course filtration shall be located in the AHU just after the outside and return air streams enter the recirculation unit. Level 1 filtration is the lowest efficiency, lowest cost and is used to remove all large particles (3.0 microns and larger such as insects and vegetation) found in outside air. The intention is to keep the internal components, (coils, fans etc.) and the AHU internal surfaces clean over an extended period. They also act as a pre-filtration for the Level 2 filtration and extend their life. A minimum of EN G4 (MERV 7) filters are recommended for Level 1.

Filter face air velocities shall not exceed of 2.5m/s (450 fpm). At the AHU maximum air volume flow rate, the initial pressure drop (clean) across the filter shall not exceed 100 Pa and the final pressure drops (dirty) should not exceed 250 Pa (1.0" w.g.) for panel/pleat filters as guidance, before completing the LCC analysis.

Fine / SECONDARY Filtration

Secondary or Fine filtration is more expensive and should be located as the last component before discharge from the AHU. This is recommended to ensure any particles or other matter (mold) generated in the AHU is captured before discharge to the ductwork and also to extend the life of filters further downstream. EN F8 or F9 (MERV 14/15/16) filters are recommended for Level 2.

Filter face air velocities shall not exceed of 2.5m/s (450 fpm). At the AHU maximum air volume flow rate, the initial pressure drop (clean) across the filter shall not exceed 100 Pa and the final pressure drops (dirty) should not exceed 450 Pa (1.4" w.g.) for panel/pleat filters as guidance, before completing the LCC analysis.Particle count test

Test covers verification of cleanliness. Dust particle counts to be carried out and result printed. The number of readings and positions of tests should be defined in accordance with ISO 14644-1 Annex B5

FILTER LEAKAGE TESTS

To verify filter integrity. Filter penetration tests to be carried out by a competent person to demonstrate filter media, filter seal and filter frame integrity. Only required on HEPA filters. Refer to ISO 14644-3 Annex B6

CONTAINMENT LEAKAGE TEST

To verify absence of cross-contamination. Demonstrate that contaminant is maintained within a room by means of:
• airflow direction smoke tests
• room air pressures.
Refer to ISO 14644-3 Annex B4

Air pressure differential

This test is used to verify 'non cross-contamination'- positive air pressure pushes out particles from clean zone. Log of pressure differential readings to be produced or critical plants should be logged daily, preferably continuously. A 15 Pa pressure differential between different zones is recommended. Refer to ISO 14644-3 Annex B5

air flow volume

To verify air change rates. Airflow readings for supply air and return air grilles to be measured and air change rates to be calculated. Refer to ISO 14644-3 Annex B13

air flow velocity

To verify unidirectional flow or containment conditions. Air velocities for containment systems and unidirectional flow protection systems to be measured. Refer to ISO 14644-3 Annex B4

RECOVERY

To verify clean-up time. Test to establish time that a cleanroom takes to recover from a contaminated condition to the specified cleanroom condition. Should not take more than 15 minutes. Refer to ISO 14644-3 Annex B13

AIR FLOW VISUALISATION

To verify required airflow patterns. Tests to demonstrate air flows:
- from clean to dirty areas
- do not cause cross-contamination
- uniformly from unidirectional airflow units

Demonstrated by actual or video-taped smoke tests. Refer to ISO 14644-3 Annex B7

9.0   UTILITY GASES & WATER

The key utilities involved for cleaning include utilities such as water, compressed gases (air, nitrogen etc.) and the heating and cooling of process equipment. Water quality can impact the effectiveness of pre-rinsing, washing, and final rinsing. Therefore, both the water temperature and quality need to be tightly controlled and monitored. Gases are typically used in order to blowdown or blowout remaining fluids or they are used as a drying step.

The term "clean utilities" in the life science industry refers to utilities that have to fulfil quality regulatory requirements. The basis of these requirements is due to the application of the utilities in the production of products or if water (e.g. water for injection) is used in the final product, or cleaning or processing where product contact may occur.

The most common utility is water, which can be supplied in different pharmaceutical grades of purity. Purified water (PW or PUW), Highly Purified Water (HPW) and Water-for-Injection (WFI) are the most common. Water quality specifications can be found in the pharmacopeias, e.g. the US Pharmacopeia. Other clean utilities can also include clean compressed air, clean gasses (e.g. nitrogen, argon and oxygen), and clean steam.

Pure steam is used in pharma and biotech for sterile application, for autoclave sterilisation etc. Distribution piping of clean steam is a critical aspect. Improper sizing of pipes may impact the production process and lead to a loss of time during sterilisation. Clean steam, also referred to as "pure steam", and gases used in manufacturing operations must be of a quality suitable for their intended purpose. The intended use of clean steam and gases must be understood in order to determine any risks to the patient or product. For example, gases that end up being part of the product must fulfil the regulatory requirements. Preventative maintenance and ongoing monitoring must be implemented for clean steam systems.

Water systems for purified water, de-ionised water and Water-for-Injection (WFI) must provide a consistent and reproducible output. Where there is moisture, there is always a risk of microbial contamination. Therefore, the design of water systems should mitigate against such risks. Good engineering practices such as using circulation loops, no dead legs and polished-surface finishes all work to provide an effective and safe system. The design should also take into account ease of sampling at the point of use. The removal of endotoxins is a requirement for WFI. Ongoing sampling monitoring the quality of water is particularly important where water systems are concerned. Procedures should be in place to ensure that effective monitoring and testing is maintained. Action limits and acceptance criteria should be clearly documented in approved SOPs or the equivalent. Failure to meet limits or acceptance criteria should initiate an investigation.

OQ Testing

Operational qualification or OQ is a formal validation activity, and as such should be completed per an approved protocol. The purpose of OQ testing is to confirm the operational and functionality of the clean steam system. This should demonstrate that all critical aspects of a URS are fulfilled. OQ verifications include:

o   Testing of temperatures and operating pressures
o   Capacity testing (under load)
o   Steam trap operation
o   Verification of automated functions and alarms
o   Check of automation systems
o   Correct function of valves and sampling points

PQ Testing

Due to the high operating temperatures and the associated lethality, clean steam systems are resistant to microbiological contamination.

Issues that arise can normally be attributed to equipment failures with the steam generator or contaminated water being supplied to the system. Bacterial endotoxin testing is used to monitor clean steam systems for both PQ purposes and throughout the life cycle of the equipment operation. Steam is condensed, sampled and tested. The condensate should meet WFI specifications with the exception of viable total aerobic count. Clean steam PQs are commonly completed using a three-phase approach to testing. The first phase ensures the system consistently operates within the required ranges and the steam provided meets the acceptance criteria. Typically, phase one bacterial endotoxin testing and physio-chemical testing is completed over a two-week period. For phase two, the same frequency and type of testing may be applied for an additional two weeks. After phase two testing, the system may be available for general use if allowed for within internal company procedures. Phase two testing at PQ should also provide a report with all results documented and reviewed. Phase three of PQ is intended to demonstrate the effective and consistent operation of the system over a longer term (approx. 12 months). Sampling is typically performed weekly.

Further Reading on Clean Steam

o   PIC/S PI009-3 – Pharmaceutical Inspection Co-Operation Scheme - Inspection of Utilities
o   EN 285 – European Standard - Sterilisation, Steam Sterilisers, Large Sterilisers
o   USP <1231> – United States Pharmacopeia - <1231> "Water for Pharmaceutical Purposes"
o   USP– United States Pharmacopeia - Monograph "Pure Steam"
o   EN 285 – European Standard - Sterilisation, Steam Sterilisers, Large Sterilisers

Schematic representation (simplified) of clean steam

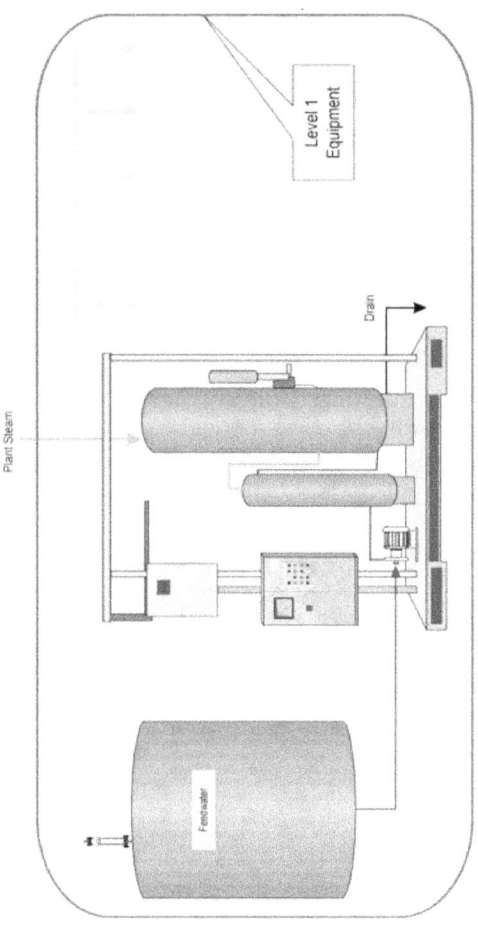

RO Water, di water and Water for Injection

Water Systems

Water supply and the associated water systems in biotechnology and pharmaceutical plants are often critical utilities and therefore, critical to quality and safety or product. Purified water is commonly used to clean equipment and vessels, to cool or heat processing pipes and systems, in sterilize products or components via moist heat sterilsiation or indeed are used to the formulation of producing the finished product (e.g. water-for-injection). Various grades of water service a particular purpose. Some common types include:

- o   RO water
- o   DI Water
- o   Purified water
- o   Water-for injection

Reverse Osmosis, RO water and Deionized, DI water are both types of purified water. They are however, produced via different processes and therefore have different characteristics.RO water is produced by forcing water through a semi-permeable membrane. Water molecules can pass through the membrane material while larger molecules are due to their size are prevented from traversing the membrane. The RO process removes many impurities, but some dissolved solids or impurities or contaminants may still be present in the water after the RO process. DI water is created by passing water through an ion exchange membrane or material that that removes charged ions by the deionization process. Therefore, DI water seen as a more purer water that is suitable for applications in pharmaceutical manufacturing where no impurities or contaminations are desired. Critical Process Parameters for a water system include:

- o   Pressure
- o   pH
- o   Conductivity Level
- o   TOC
- o   Flow rate
- o   Temperature
- o   Resistivity

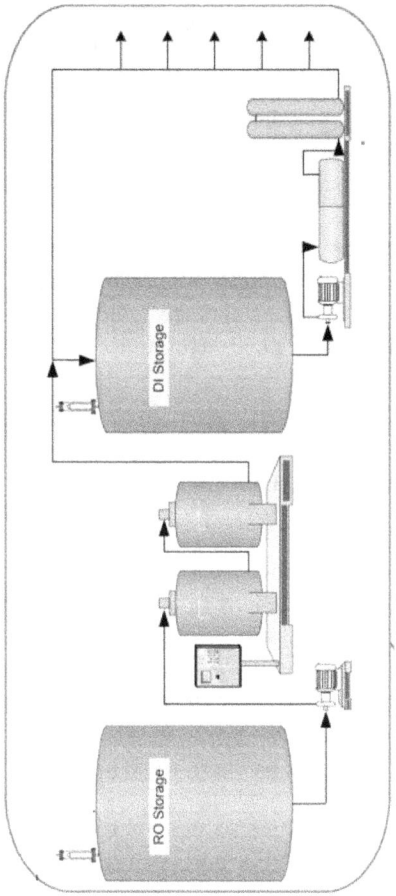

The schematic above shows the standard list of sub-systems and equipment for RO or DI water. The potential CPPs are listed above. CPPs are needed to ensure the system produces the desired quality:

The CQAs and CPPs are routinely monitored through the calibrated monitoring system which ensures any equipment failures would be detected.

- o Pressure
- o pH
- o Conductivity Level
- o TOC
- o Flow rate
- o Temperature
- o Resistivity

Water for Injection:

WFI is sterile and pyrogen-free water containing no less than 10 CFU/100ml (Colony Forming Units) with a sample size of between 100 and 300 ml and an endotoxin level < 0.25 EU/ml. The use of WFI is two-fold. Firstly, it can be used for critical processing steps such as washing and rinsing. It can also be used in injectable products. WFI is a key raw material for sterile intravenous and intradermal products. WFI is produced by a Multi-Column Distillation Plant (MCDP) and must meet the microbial requirements of regulated bodies.

The cleaning of equipment, vessels and process piping is a critical activity. Any residue from a previous production batch needs to be removed in order to avoid cross-contamination. Clean in Place and Sterilize in Place skids are often utilised to allow efficient switchover between batches and/or products. Where possible, manual cleaning should be avoided unless essential due to the design of a system or particular location or configuration.

Compressed Air- Generation, storage and distribution

Compressed air is used for valve actuation, instrument air and process air to name but a few applications. Only the point-of-use filtration and the gas quality instrumentation should be classified as level 1. When flow or pressure is a CPP, the measurement/monitoring should be performed by the system into which the gas is flowing. Additionally, the CQAs and CPPs should be routinely monitored through the calibrated monitoring system. For compressed air, the potential CPPs are listed below. For the physical system being evaluated, the use and the application of the compressed air will determine which (if not all) CPPs are needed to ensure the system produces product of the desired quality.

- ➢ Hydrocarbons
- ➢ Moisture
- ➢ Particulates
- ➢ Temperature

It is important that each point of use has appropriate sterile filters in place. If the filter is not placed directly at the point of use, control and counter measures should be implemented to address any risk of contamination downstream of the filter. Compressed air for bio-pharmaceutical use must be generated using oil free compressors with appropriate temperature controls in place.

Requirement	Clean Compressed Air (impacts product quality)	Sterile Compressed Air (impacts sterility of product)
Oil content	*Not great than 0.1mg/m^3 (ISO 8573-1 Class 2)	
Microbiological requirement	Meets requirements of the environmental zone served (e.g. ISO 5)	Sterile
Filtration	Minimum 0.45µm	0.2µm membrane

requirement	membrane filter	filter

Compressed Air Design Requirements

Compressed Air generation systems are required to address the following components in order to produce compressed air that complies to ISO 8573-1 requirements.

Class	ISO 8573-1					Viable particle counts by Air sampling Method
	Solid Particulate			Water content	Oil content	
	Maximum no. of particle per m³			Vapor pressure Dew point	Total oil mg/ m³	
	0.1-0.5 µ	0.5-1µ	1-5µ			
0	As specified by the user / supplier (≤ class 1)					
1	100	1	0	-70°C	0.01	100 CFU/m³
2	100,000	1000	10	-40°C	0.1	
3	-	10,000	500	-20°C	1	

To remind oneself of the qualification process, refer to Chapter on Qualification Requirements which specifies the qualification expectations for medicinal products. It is also an approach that can be adopted by other industries of disciplines within life sciences or Medtech.

Design Qualification

So far, in relation to compressed air generation, distribution and storage we have summarized the design elements and design requirements. (Previous sections). Both the design elements and requirements are inputs to the Design Qualification. At this point in a project an approved User requirements specification should also be available. DQ is an evaluation of the design elements and design requirements that the URS and Vendor specifications. Note: Vendor specifications are often documented in a Functional design specification (FDS) which in simple terms is an 'answer' to each of the requirements specified in the URS.

DQ Evaluation

In this section the design requirements are benchmarked against the URS and Vendor responses or specifications. The Description is based on the design element and design requirement. The URS requirements are assumed to be already approved in the separate document. The vendor specification is the response of the vendor and can be a specific document that is created or alternatively if oof-the-shelf it may be a operating manual of similar document.

Description: Capacity

- O User Requirements Specification: Generation of 1400 CFM with outlet pressure of 6-8 kg/cm2.

- O Vendor Specification: Generation of 1600 cfm with outlet pressure of 6-8 kg/cm2.

- O Verification: While the vendor specified system has a higher capacity, this is acceptable.

Description: Inlet air filtration

o   User Requirements Specification: 3 microns with 99% efficiency
o   Vendor Specification: 3 microns with 99% efficiency
o   Verification: Requirement is met by design and vendor.

Description: Compressed air generation

o User Requirements Specification: Screw, non-lubricated oil free, air cooled.
o Vendor Specification: screw, non-lubricated oil free, air cooled.
o Verification: Require met by vendor

Description: Inter cooler
o User Requirements Specification: Air or water cooled
o Vendor Specification: air cooled type
o   Verification: Air cooled type is acceptable.

Description: After cooler

o User Requirements Specification: Air or water cooled
o Vendor Specification: air cooled type
o   Verification: Air cooled type is acceptable.

Description: Dryer

o User Requirements Specification: Must be inbuilt to the compressor unit with dryer to produce dew point -20DegC or better per ISO 8573
o Vendor Specification: Generation of 1600 cfm with outlet pressure of 6-8 kg/cm2.
o Verification: as specified above

This process is then replicated for the remaining user requirements. A successful DQ review will ensure all design aspects are review with accepable vendor responses to the user requirements and design intent.

www.ingramcontent.com/pod-product-compliance
Lightning Source LLC
Chambersburg PA
CBHW080951290526
45795CB00009B/2953